*How Ghee Ang and
Sreekumar Pisharath*

Energetic Polymers

Related Titles

Fink, J. K.

Handbook of Engineering and Specialty Thermoplastics

Volume 1: Polyolefins and Styrenics

2010
E-Book
ISBN: 978-1-118-02928-2

Lackner, M., Winter, F., Agarwal, A. K. (eds.)

Handbook of Combustion

2010
ISBN: 978-3-527-32449-1

Leclerc, M., Morin, J.-F. (eds.)

Design and Synthesis of Conjugated Polymers

2010
ISBN: 978-3-527-32474-3

Agrawal, J. P.

High Energy Materials

Propellants, Explosives and Pyrotechnics

2010
ISBN: 978-3-527-32610-5

Pascault, J.-P., Williams, R. J. J. (eds.)

Epoxy Polymers

New Materials and Innovations

2010
ISBN: 978-3-527-32480-4

Menczel, J. D., Prime, R. B.

Thermal Analysis of Polymers, Fundamentals and Applications

2009
E-Book
ISBN: 978-0-470-42377-6

How Ghee Ang and Sreekumar Pisharath

Energetic Polymers

Binders and Plasticizers for Enhancing Performance

WILEY-VCH Verlag GmbH & Co. KGaA

The Authors

Prof. How Ghee Ang
Nanyang Technological Univ.
Energetics Research Institute
50 Nanyang Av. Block N1-B4a-02
Singapore 639798
Singapore

Dr. Sreekumar Pisharath
Nanyang Technological Univ.
Energetics Research Institute
50 Nanyang Av. Block N1-B4a-02
Singapore 639798
Singapore

All books published by **Wiley-VCH** are carefully produced. Nevertheless, authors, editors, and publisher do not warrant the information contained in these books, including this book, to be free of errors. Readers are advised to keep in mind that statements, data, illustrations, procedural details or other items may inadvertently be inaccurate.

Library of Congress Card No.: applied for

British Library Cataloguing-in-Publication Data
A catalogue record for this book is available from the British Library.

Bibliographic information published by the Deutsche Nationalbibliothek
The Deutsche Nationalbibliothek lists this publication in the Deutsche Nationalbibliografie; detailed bibliographic data are available on the Internet at <http://dnb.d-nb.de>.

© 2012 Wiley-VCH Verlag & Co. KGaA, Boschstr. 12, 69469 Weinheim, Germany

All rights reserved (including those of translation into other languages). No part of this book may be reproduced in any form – by photoprinting, microfilm, or any other means – nor transmitted or translated into a machine language without written permission from the publishers. Registered names, trademarks, etc. used in this book, even when not specifically marked as such, are not to be considered unprotected by law.

Print ISBN: 978-3-527-33155-0
Typesetting MPS Limited, a Macmillan Company, Chennai, India
Printing and Binding Markono Print Media Pte Ltd, Singapore
Cover Design Formgeber, Eppelheim

Contents

Preface *IX*
Abbreviations *XI*

1	**Polymers as Binders and Plasticizers – Historical Perspective** *1*	
1.1	Nitrocellulose *1*	
1.2	Polysulfides *2*	
1.3	Polybutadienes (PBAA, PBAN, and CTPB) *3*	
1.4	Polyurethanes *4*	
1.5	Hydroxy Terminated Polybutadiene *5*	
1.6	Explosive Binders *5*	
1.7	Thermoplastic Elastomers *6*	
1.8	Energetic Polymers (Other Than NC) as Binders *8*	
1.8.1	Polyglycidyl Nitrate *8*	
1.8.2	GAP *9*	
1.8.3	Energetic Polyoxetanes *10*	
1.8.4	Polyphosphazenes *11*	
1.8.5	Energetic Thermoplastic Elastomers *11*	
1.9	Energetic Polymer Plasticizers *12*	
	References *15*	
2	**High Nitrogen Content Polymers** *19*	
2.1	Introduction *19*	
2.2	Preparation of Energetic Azido Polymers *19*	
2.2.1	Glycidyl Azide Polymer *19*	
2.2.2	Azido Polymers from Oxetanes *22*	
2.2.2.1	Poly(BAMO) *23*	
2.2.2.2	Poly(AMMO) *24*	
2.3	Physical Properties of Azido Polymers *25*	
2.4	Curing of Azido Polymers *25*	
2.4.1	Curing by Polyisocyanates *25*	

Energetic Polymers: Binders and Plasticizers for Enhancing Performance, First Edition.
How Ghee Ang and Sreekumar Pisharath.
© 2012 WILEY-VCH Verlag GmbH & Co. KGaA, Weinheim. Published 2012 by WILEY-VCH Verlag GmbH & Co. KGaA

2.4.1.1	Preparation 26
2.4.1.2	Structure of Cured Polyurethane Elastomer 28
2.4.1.3	Kinetics of Curing of Azido Polymers 29
2.4.1.4	Gel-Time Characteristics 31
2.4.1.5	Post-Cure Properties 33
2.5	Curing of Azido Polymers by Dipolarophiles 38
2.6	Thermal Decomposition Characteristics of Azido Polymers 41
2.6.1	Mechanism of Thermal Decomposition 41
2.6.2	Kinetics of Thermal Decomposition of Azido Polymers 43
2.7	Combustion of Azido Polymers 47
2.8	Thermal Decomposition and Combustion of Energetic Formulations with Azido Polymers 49
2.8.1	Thermal Decomposition of Azido Polymer/Nitramine Mixtures 50
2.8.2	Combustion of Azido Polymer/Nitramine Propellant Mixtures 54
2.8.3	Combustion of Azido Polymer/Ammonium Nitrate Composite Propellants 57
2.8.4	Combustion of Azido Polymer Propellants with HNF 58
2.9	Performance of Azido Polymer-Based Propellants 61
2.10	Azido Polymers as Explosive Binders 63
2.10.1	Azido Polymer Based PBX Formulations for Underwater Explosives 65
2.11	Tetrazole Polymers and Their Salts 66
2.12	N–N-Bonded Epoxy Binders 69
	References 70

3 Nitropolymers as Energetic Binders 81

3.1	Introduction 81
3.2	Preparation of Nitropolymers 82
3.2.1	Nitrocellulose 82
3.2.2	Poly(Glycidyl Nitrate) (PGN) 83
3.2.2.1	Curing of PGN 85
3.2.3	Poly(Nitratomethyl-methyl Oxetane) (Poly(NIMMO)) 88
3.2.4	Nitrated HTPB (NHTPB) 90
3.2.5	Nitrated Cyclodextrin Polymers (Poly(CDN)) 90
3.3	Thermal Decomposition Behavior of Nitropolymers 91
3.3.1	Nitrocellulose 91
3.3.2	Poly(Nitratomethyl-Methyl Oxetane) (Poly(NIMMO)) 94
3.3.3	Poly(Glycidyl Nitrate) (PGN) 98
3.4	Combustion of Nitrate Ester Polymers and Propellants 100
3.4.1	NC Based Double-Base Propellants 100
3.4.2	Composite Modified Double-Base (CMDB) Propellants 101
3.4.3	Poly(NIMMO)-Based Composite Propellants 106

3.4.4	PGN-Based Composite Propellants 109
3.5	Nitropolymer-Based Explosive Compositions 111
	References 115

4	**Energetic Thermoplastic Elastomers** 121
4.1	Introduction 121
4.2	Preparation of Energetic Thermoplastic Elastomers 122
4.3	Thermal Decomposition and Combustion of ETPEs 132
4.4	Combustion of ETPE Propellant Formulations 137
4.5	Performance of ETPE Based Propellant Formulations 139
4.6	Melt-Cast Explosives Based on ETPEs 141
4.7	ETPE Based Polymer Nanocomposites 142
	References 144

5	**Fluoropolymers as Binders** 147
5.1	Introduction 147
5.2	Poly(tetrafluoroethylene) (PTFE) 148
5.2.1	Phase Transitions of PTFE 148
5.2.2	Energetic Composites of PTFE 149
5.3	Copolymers of Tetrafluoroethylene 152
5.3.1	Kel-F800 153
5.3.1.1	Dynamic Behavior of Kel-F800 under Shock 153
5.3.1.2	Thermal Decomposition 155
5.3.2	Viton A 156
5.3.2.1	Thermal Decomposition of Viton 158
5.4	Energetic Polymers Containing Fluorine 159
5.5	Miscellaneous Energetic Fluoropolymers 161
5.6	PBX Formulations with Fluoropolymers 162
	References 165

6	**Energetic Plasticizers for High Performance** 171
6.1	Introduction 171
6.2	Energetic Plasticizers Based on Azido Compounds 172
6.2.1	Azido Acetate Ester Based Plasticizers 172
6.2.2	Azido Based Oligomeric Plasticizers 174
6.3	Performance of Propellant Formulations Containing Azido Plasticizers 178
6.4	Nitrate Ester Plasticizers 180
6.4.1	General Characteristics 180
6.4.2	Performance 183
6.5	Nitrate Ester Oligomers as Energetic Plasticizers 184
6.6	Miscellaneous Plasticizers Based on Nitro-Groups 186
6.6.1	Polynitro-Aliphatic Plasticizers 186

6.6.2	Nitratoethylnitramine (NENA) Plasticizers *188*	
	References *189*	
7	**Application of Computational Techniques to Energetic Polymers and Formulations** *193*	
7.1	Introduction *193*	
7.2	Overview of Computational Techniques *193*	
7.3	Application of Computational Modeling Techniques to Energetic Polymer Formulations *197*	
7.3.1	Quantum Mechanical Methods *197*	
7.3.2	Molecular Dynamics (MD) Simulations *201*	
7.3.3	Mesoscale Simulations *205*	
7.3.4	Macroscale Simulations *206*	
7.4	Future Outlook *207*	
	References *208*	

Index *211*

Preface

Over the years, a variety of polymer binders have been developed and used in energetic material composite applications to cope with the global demand for insensitive and high performance propellant and explosive formulations. One of the important milestones in this effort was the development of polymers containing energetic groups, namely azido or nitro groups that release significant amounts of energy under the application of a thermal stimulus. Such type of polymers are referred as energetic polymers. Energetic polymers, when used as binders, provide additional energy and insensitivity to formulations as compared with their inert counterparts. Low molecular weight energetic polymers, with the advantages of a lower migration tendency and less sensitivity to external stimuli, are finding application as energetic plasticizers. Through the years, a range of energetic polymers have been developed as prospective candidates for application as binders and plasticizers. Some of them, namely GAP, Poly(NIMMO), and Poly(GLYN), have reached the stages of mass production.

This book is intended to provide a state-of-the-art overview of the various energetic polymers employed for propellant and explosive formulations. The book is divided into seven chapters. The first chapter details the historical development of the emergence of polymers, including that of the energetic ones, as a vital component in energetic material composites. Chapters 2 and 3 discuss the two important members of the energetic polymer family; the high nitrogen containing polymers and nitropolymers. For the former, the high nitrogen content (in the form of azido and tetrazole groups) provides the polymers with high positive heat of formation and superior energetic properties, albeit resulting in fuel-rich formulations. However, for the latter, the presence of nitro groups offers a better balance of oxygen content and energetic properties. The majority of the commercialized energetic polymers for binder and plasticizer applications belong to these two families. Comparisons are provided wherever possible with the inert binder formulations, to stress the capability of energetic binder formulations for enhancing performance and insensitivity.

Energetic thermoplastic elastomers (ETPEs) have evinced an immense interest due to their unique mechanical and thermal behavior. They are copolymers of energetic polymers that are discussed in Chapters 2 and 3. ETPEs provided a paradigm shift in the processing of energetic material formulations by replacing the cast curing process with an environment friendly melt casting process. Chapter 4 presents the various ETPEs used as binders and their application in propellant and explosive formulations.

The application of fluoropolymers as binders is discussed in Chapter 5. Fluoropolymers owe their reactivity to the high electronegativity and oxidizing nature of fluorine. Higher combustion energies could be obtained from composites of fluoropolymers with nanoparticles of combusting metal fuels, such as magnesium and aluminum. This possibility has opened up a new area of energetic materials research called reactive nanomaterials.

Plasticizers derived from energetic polymers with nitro and azido substituents have lower migration tendency as compared with the nitrate ester based energetic plasticizers used in explosive and propellant formulations. The preparation, physical properties, and energetics of azido/nitro polymer based energetic plasticizers are discussed in Chapter 6.

The macroscopic behavior of energetic material composites is the result of the complicated physical processes and microstructure occurring in the composite at multiple length and time scales. Therefore, a multiscale simulation approach is necessary for the prediction of macroscopic properties of energetic material composites from fundamental molecular processes. Simulation of such systems requires theoretical models that range from those including atomistic effects to macroscopic continuum models and passing through the intermediate mesoscopic simulations. Chapter 7 introduces the concept of multiscale modeling in energetic composites and application to simulate the physical and chemical processes in the energetic composites. We hope that the energetic polymers and formulations presented here will be useful in enhancing the search for new high performance and insensitive energetic material composites.

Prof. How Ghee Ang
Dr. Sreekumar Pisharath
Singapore

Abbreviations

A	Frequency Factor
AA	Acrylic acid
ABL	Allegheny Ballistics Laboratory
AC	Acrolyl chloride
ACE	Activated Chain End
ADN	Ammonium dinitramide
ADDF	2,2-Dinitro-1,3-propanediol diformate
AFM	Atomic Force Microscopy
AFEM	Atomic-Scale Finite-Element Method
AIAA	American Institute of Aeronautics and Astronautics
AMM	Activated Monomer Mechanism
AMMO	3-Azidomethyl 3-methyl oxetane
AN	Ammonium nitrate
AP	Ammonium perchlorate
ARC	Adiabatic Reaction Calorimetry
ARL	Army Research Laboratory, USA
ASME	American Society of Mechanical Engineers
AWE	Atomic Weapons Establishment, UK
BABAMP	Bis(azido acetoxy) bis(azido methyl) propane
BAMO	3,3-Bis(azidomethyl) oxetane
BBrMO	3,3-Bis(bromomethyl) oxetane
BCMO	3,3-Bis(chloromethyl) oxetane
BDO	1,4-Butanediol
BDNPA	Bis(2,2-dinitropropyl) acetal
BDNPF	Bis(2,2-dinitropropyl) formal
BDNPA/F	Bis(2,2-dinitropropyl) acetal/formal
BE	Boundary Element
BTTN	1,2,4-Butanetriol trinitrate
BTATz	3,6-Bis(1H-1,2,3,4-tetrazol-5-ylamino)-s-tetrazine
Bu-NENA	Butyl-N-(2-nitratoethylnitramine
B3LYP	Becke 3-Parameter (Exchange) Lee Yang Parr
CAB	Cellulose acetate butyrate
CAL	Calorie

Energetic Polymers: Binders and Plasticizers for Enhancing Performance, First Edition.
How Ghee Ang and Sreekumar Pisharath.
© 2012 WILEY-VCH Verlag GmbH & Co. KGaA, Weinheim. Published 2012 by WILEY-VCH Verlag GmbH & Co. KGaA

CC	Copper chromite
CD	Cyclodextrin
CDN	Nitrated cyclodextrin
CEP	Tris(β-chloroethyl) phosphate
CFC	Chlorofluorocarbon
CG	Coarse Grain
CL-20	2,4,6,8,10,12-Hexanitro-2,4,6,8,10,12-hexaazaisowurtzitane (HNIW)
ClMMO	3-Chloromethyl-3-methyl oxetane
CMDB	Composite Modified Double Base
Comp. B	Explosive composition based on RDX and TNT
COMPASS	Condensed-phase Optimized Molecular Potentials for Atomistic Simulation Studies
CPX-413	UK's extremely insensitive explosive composition based on NTO, HMX, and Poly(NIMMO).
CRP	Controlled Radical Polymerization
CTPB	Carboxy terminated polybutadiene
CTFE	Chlorotrifluoroethylene
DAAF	4,4-Diamino-3,3-azoxyfurazan
DB	Double Base
DBTDL	Dibutyltin dilaurate
DBP	Dibutyl phthalate
p-DCC	p-Bis (α,α-dimethylchloromethyl) benzene
DDM	4,4′-Diaminodiphenylmethane
DEGBAA	Diethylene glycol bis-azido acetate
DEGDN	Diethylene glycol dinitrate
DHT	3,6-Dihydrazino-s-tetrazine
DMA	Dynamic Mechanical Analysis
DMSO	Dimethyl sulfoxide
DMF	Dimethyl formamide
DNAN	2,4-Dinitroanisole
DPD	Dissipative Particle Dynamics
DOS	Dioctyl sebacate
DREV	Defense Research Establishment Valcartier, Canada
DSC	Differential Scanning Calorimetry
DSTO	Defence Science and Technology Organization
DTA	Differential Thermal Analysis
DTG	Derivative Thermogravimetric Analysis
DTIC	Defense Technical Information Center
DRA	Defense Research Agency, UK
E_A	Activation energy
ECH	Epichlorohydrin
EGBAA	Ethylene glycol bis-azido acetate
EGDN	Ethylene glyclol dinitrate
EIDS	Extremely Insensitive Detonating Substance

EOS	Equation of State
ETPE	Energetic Thermoplastic Elastomer
Et-NENA	Ethyl-N-nitratoethylnitramine
ESD	Electrostatic Discharge
ESA	European Space Agency
ESR	Electron Spin Resonance
ESTANE	Thermoplastic elastomer based on segmented polyurethane
FOI	Swedish Defence Research Agency
F of I	Figure of Insensitivity
FOX-7	1,1-Diamino-2,2-dinitroethylene
FK-800	A fluorocopolymer developed by 3M having equivalent composition to that of Kel-F800
FTIR	Fourier Transform Infrared
GA	Glycidyl azide
GAP	Glycidyl azide polymer
GAPA	Azido terminated GAP
GC	Gas Chromatography
GLyN (GN)	Glycidyl nitrate
GPC	Gel Permeation Chromatograph
GZT	Guanidine-5,5′-azotetrazolate
H-6	Underwater explosive composition developed by Australia
HDPE	High Density Polyethylene
HFP	Hexafluoropropylene
HFC	Heat Flow Calorimetry
HMMO	3-Hydroxy methyl-3-methyl oxetane
$H_{12}MDI$	4,4′-Dicyclohexyl methyl diisocyanate (4,4′-diphenylmethane diisocyanate)
HMX	High Melting Explosive or Her Majesty's Explosive
HNF	Hydrazinium nitroformate
HOMO	Highest Occupied Molecular Orbital
HTPB	Hydroxy Terminated Polybutadiene
HYFLON	Fully fluorinated semicrystalline copolymer of tetrafluoroethylene (TFE) and 2,2,4-trifluoro-5-trifluoro-methoxy-1,3-dioxide (TTD)
ICE	Isentropic Compression Experiment
ICI	Imperial Chemical Industries
ICT	Fraunhofer Institut Chemische Technologie, Germany
IM	Insensitive Munition
IMEMTS	Insensitive Munitions and Energetic Materials Technical Symposium
IPDI	Isophorone diisocyanate
I_{sp}	Specific Impulse
IDP	Isodecyl pelargonate
J	Joule
JANNAF	Joint Army–Navy–NASA–Air Force
JA2	Nitrocellulose based propellant

JAX	JA2 propellant loaded with RDX
K	Kelvin
K-10	A mixture of 2-nitroethylbenzene and 2,4,6-trinitroethylbenzene
Kel-F800	Copolymer of vinylidene and hexafluoropropylene
Kraton	Copolymer of styrene and ethylene/butylenes
LANL	Los Alamos National Laboratory
LLNL	Lawrence Livermore National Laboratory
LLM-105	2,6-Diamino-3,5-dinitropyrazine-1-oxide
LOVA	Low Vulnerability Ammunition
LUMO	Lowest Unoccupied Molecular Orbital
LX	Series of explosive formulations developed by LLNL based on fluoropolymer binders
MAPO	Tris(1-(2-methylaziridinyl) phosphine oxide)
MCHI	Methylene bis(cyclohexyl isocyanate)
MD	Molecular Dynamics
MHMO	3-Methyl 3-hydroxymethyl oxetane
MIL-STD	Military Standard
MMA	Methyl methacrylate
M_c	Average molecular weight between crosslinks
M_n	Number average molecular weight
M_w	Weight average molecular weight
Me-NENA	Methyl-N-(2-nitroxyethyl) nitramine
MODPOT	*Ab initio* Core Model Potential
MOVO	Molybednum/vanadium oxide catalysts
MRD-CI	Multiple Reference Double Excitation–Configuration Interaction
MS	Mass Spectroscopy
MSIAC	Munitions Safety Information Analysis Center
MTH	Mathematical Theory of Homogenization
MTMO	3-Methyl-3'-(tosyloxymethyl) oxetane
NASA	National Aeronautic Space Agency
NATO	North Atlantic Treaty Organization
NC	Nitrocellulose
NCO	Isocyanate
NDIA	National Defence Industrial Association
NENA	N-(2-nitroxyethyl) nitramine (nitratoethylnitramine)
NG	Nitroglycerin
NHTPB	Nitrated hydroxy terminated polybutadiene
NIMMO	3-Nitratomethyl 3-methyl oxetane
NIMIC	NATO Insensitive Munitions Information Center, USA
NMR	Nuclear Magnetic Resonance
NSWC	Naval Surface Warfare Centre
NTO	3-Nitro-1,2,4-triazole-5-one
OB	Oxygen Balance
ODTX	One Dimensional Time to Explosion
OM	Optical Microscopy

ONR	Office of Naval Research
P1	1,3-Bis(azidoacetoxy)-2-azidoacetoxymethyl-2-ethylpropane
PEAA	1,3-Bis(azidoacetoxy)-2,2-bis(azidomethyl) propane
PAN	Polyacrylonitrile
PAX	Picatinny Arsenal Explosive
PB	Polybutadiene
PBA	Poly(butylene adipate)
PBAA	Poly(butadiene–acrylic acid)
PBAN	Poly(butadiene–acrylic acid–acrylonitrile)
PBX	Polymer Bonded Explosive
PBXW-115	PBX developed in USA for underwater applications
PBXN	PBXs developed by US Navy
P_{CJ}	Chapman–Jouguet Pressure
PCTFE	Polychlorotrifluoroethylene
PDMS	Poly(dimethyl siloxane)
PECH	Poly(epichlorohydrin)
PEG	Polyethylene glycol
PETKAA	Pentaerythritol tetrakis (azidoacetate)
PMMA	Poly(methyl methacrylate)
PMVT	Poly(2-methyl-5-vinyl tetrazole)
PNC	Polymer Nanocomposite
Poly(AMMO)	Poly(3-azidomethyl 3-methyloxetane)
Poly(BAMO)	Poly(3,3-bis(azidomethyl) oxetane)
Poly(BEMO)	Poly(bis(ethoxymethyl) oxetane)
Poly(CDN)	Nitrated cyclodextrin polymers
Poly(GlyN) or PGN	Poly(glycidyl nitrate)
Poly(NiMMO)	Poly(3-nitratomethyl 3-methyl oxetane)
Pr-NENA	Propyl-N-(2-nitroxyethyl) nitramine
Pe-NENA	Pentyl-N-(2-nitroxyethyl) nitramine
PS	Poly(styrene)
PTC	Phase Transfer Catalyst
PTFE	Poly(tetrafluoroethylene)
PU	Polyurethane
PVT	Poly(vinyl tetrazole)
PVDF	Poly(vinylidene difluoride)
QM	Quantum Mechanics
QRX077	Explosive formulation containing HMX, FOX-7, and inert binder
RDX	Research Department Explosive
RVE	Representative Volume Element
SBAT	Simulated Bulk Auto-Ignition Temperature
SEM	Scanning Electron Micrograph
SMATCH	Simultaneous Mass and Temperature Change
SNPE	Societe Nationale des Poudres et Explosifs, France
T	Absolute Temperature

T_g	Glass Transition Temperature
T_m	Melting Temperature
TAAMP	Tris(azido acetoxy methyl) propane
TATB	1,3,5-Triamino-2,4,6-trinitrobenzene
TDI	Toluene diisocyanate
TEGDN	Triethylene glycol dinitrate
TEX	4,10-Dinitro-2,6,8,12-tetraoxa-4,10-diazaisowurtzitane
TFE	Tetrafluoroethylene
TGA	Thermogravimetric Analysis
THF	Tetrahydrofuran
TMETN	1,1,1-Trimethylol ethane trinitrate
TMNTA	Trimethylol nitromethane tris(azido acetate)
TMP	Trimethylol propane
TNT	Trinitrotoluene
TNAZ	1,3,3-Trinitroazetidine
TPB	Triphenyl bismuth
TPE	Thermoplastic Elastomer
TR	Technical Report
TTD	2,2,4-Trifluoro-5-trifluoro-methoxy-1,3-dioxide
UK	United Kingdom
USA	United States of America
VDF	Vinylidene difluoride
Viton	Copolymer of vinylidene fluoride and hexafluoropropylene
VISAR	Velocity Interferometer System for Any Reflector
VOD	Velocity of Detonation
VTS	Vacuum Thermal Stability

1
Polymers as Binders and Plasticizers – Historical Perspective

Ever since the introduction of nitrocellulose (NC) as an explosive fill in the 1850s, polymers have contributed considerably to advancements in the technology of both propellants and explosives. In addition to the specific instance of the use of NC polymer in explosive fills, applications of polymers have been most extensive in binders and plasticizers. Over the years, with the maturity of composite propellant and polymer bonded explosive technology, diverse classes of polymers have been developed for binder applications, in order to meet the dual objectives of insensitivity and high performance. This chapter focuses on the historical development of polymers for propellant and explosive formulations, particularly as binders and plasticizers.

1.1
Nitrocellulose

Nitrocellulose (NC) (Figure 1.1), a nitrated carbohydrate, was the first polymer to be used in energetic material formulations, particularly in smokeless propellants. NC was discovered by Christian Friedrich Schonbein in Basel and Rudolf Christian Bottger in Frankfurt-am-Main around 1845–1847 [1].

Henri Braconnot and Theophile Jules Pelouze had unknowingly prepared NC in 1833 and 1838, respectively, in France. They named these combustible materials

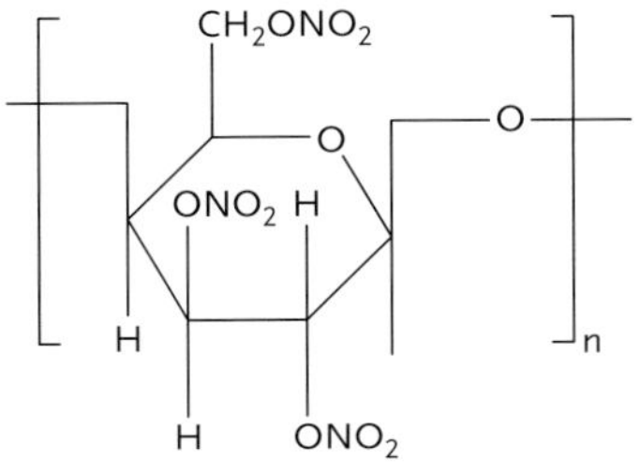

Figure 1.1 Chemical structure of nitrocellulose.

Energetic Polymers: Binders and Plasticizers for Enhancing Performance, First Edition.
How Ghee Ang and Sreekumar Pisharath.
© 2012 WILEY-VCH Verlag GmbH & Co. KGaA, Weinheim. Published 2012 by WILEY-VCH Verlag GmbH & Co. KGaA

xyloidine and nitramidine. The announcements from Schonbein and Bottger in 1846 that NC had been prepared, meant that their names have since been associated with the discovery and utilization of NC. Schonbein's process became known through the publication of an English patent to John Taylor British Patent 11407, (1846). NC was prepared by immersing cotton in a 1 : 3 mixture of nitric and sulfuric acids, which was washed in a large amount of water to remove the free acids, and then pressed to remove as much water as possible. However, NC produced by Schonbein's process was unstable, due to the rapid decomposition of the material, which was promoted by the free acid generated during the process. In 1865, Sir Frederick Abel's patent British Patent 1102, (1865). On improvements to the preparation and treatment of NC demonstrated that a pulping process could greatly increase the stability of the NC. Pulping breaks up the long fiber into shorter pieces so that the remaining acids can then be easily washed out of it. In 1868, Abel's assistant, E.A. Brown, demonstrated the first application of NC as an explosive fill, which was later employed extensively in naval mines and shells during World War II.

The application of NC as a binder was exploited when it was used for propulsion purposes in homogenous propellants, with the invention of the smokeless powder, Poudre B, by Paul Vieille in 1884 [2]. It was made by treating a mixture of soluble and insoluble NC with a 2 : 1 ether–alcohol mixture, kneading it to form a thick jelly, and rolling into thin sheets. The NC binder provided the necessary structural integrity for the propellant, which could be molded to conform to a wide range of motor geometries and be used to deliver long duration thrust. NC has been used as a single-base propellant, but only to a limited extent due its negative oxygen balance (-28.6). This was followed by the development of Ballistite (a gelatinous mixture of nitroglycerin (NG) and soluble NC in varying proportions with a small amount of aniline or diphenylamine stabilizer) invented by Alfred Nobel in 1888 and Cordite (a combination of 58% NG, 37% NC, and 5% Vaseline) by Abel and James Dewar. A mixture of NC with NG results in a higher energy propellant, not only because of the energetic nature of the NG, but also the positive oxygen balance of NG ($+3.5$) results in complete oxidation of the NC. Ballistite and Cordite are used as double and triple base propellants, which are still in widespread use [3].

1.2
Polysulfides

Polysulfide (Figure 1.2) was the first polymer to be used as a binder in the heterogeneous (composite) family of propellants in 1942 [4]. It was invented by Dr. Joseph C. Patrick in 1928 as a condensation product of ethylene dichloride

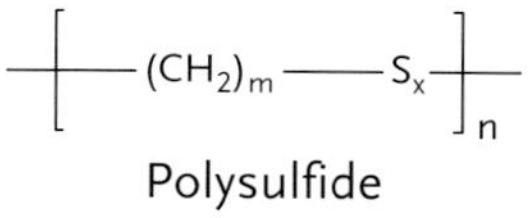

$$\left[(CH_2)_m - S_x\right]_n$$

Polysulfide

Figure 1.2 Chemical structure of polysulfide.

with sodium polysulfide. He named this polymer Thiokol and formed the Thiokol Corporation to commercialize the product. In 1945, JPL engineer Charles Bartley used polysulfide polymer (known commercially as LP-3) to formulate a new type of composite solid propellant [5, 6].

The sulfur in the polymer backbone functioned as an oxidizer in the combustion process contributing towards higher specific impulses. The polymer could be cross-linked by oxidative coupling with curatives such as p-quinonedioxime or manganese dioxide to form disulfide (–S–S–) bonds. The cured elastomers have good elongation properties for wider operating temperature ranges.

1.3 Polybutadienes (PBAA, PBAN, and CTPB)

In early 1955, the role of aluminum as a high-performance ingredient in propellant formulations was demonstrated by Charles B. Henderson's group at the Atlantic Research Corporation, USA. The polysulfide propellant developed by Thiokol could not be adapted to the use of aluminum, because chemical reactions during storage led to explosions. Therefore a new series of binders based on butadienes were developed by Thiokol. Furthermore, a polybutadiene chain polymer was found to be more favorable than a polysulfide chain to provide higher elasticity [7]. The first of the butadiene polymers to be used in a propellant was the liquid copolymer of butadiene and acrylic acid, PBAA (Figure 1.3) developed in 1954 in Huntsville, Alabama, USA.

The PBAA is prepared by the emulsion radical copolymerization of butadiene and acrylic acid. The very low viscosity of these polymers permitted the development of propellants with higher solids. However, owing to the method of preparation, the functional groups are distributed randomly over the chain. Hence, propellants prepared with PBAA show poor reproducibility of mechanical properties [8].

The mechanical behavior and storage characteristics of butadiene polymers were improved by using terpolymers based on butadiene, acrylonitrile, and acrylic acid (PBAN) (Figure 1.4) developed by Thiokol in 1954. The introduction of an acrylonitrile group improves the spacing of the carboxyl species.

This polymer has a low viscosity and low production costs. The curing systems for PBAN are based on di- or tri-functional epoxides (commercial name: Epon X-801) or aziridines (commercial name: MAPO). PBAN propellants can provide a better specific impulse, but require elevated curing temperatures [9]. The PBAN polymer was soon used in propellant formulations such as TP-H-1011, which is

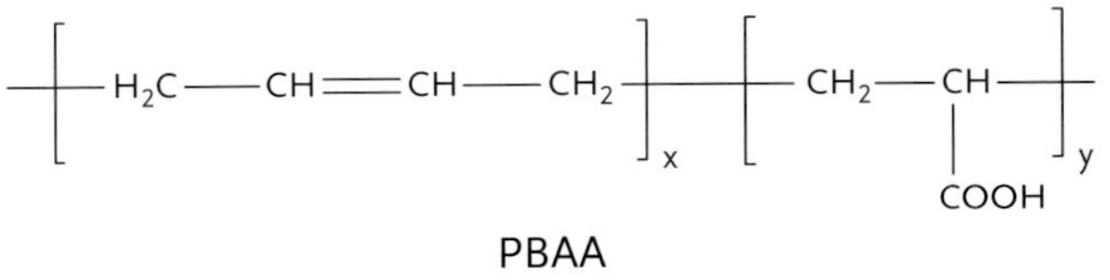

Figure 1.3 Chemical structure of PBAA.

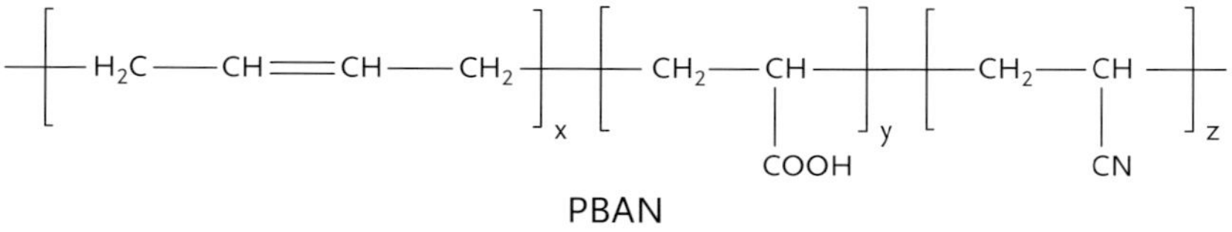

$$\left[-H_2C-CH=CH-CH_2-\right]_x \left[-CH_2-\underset{\underset{COOH}{|}}{CH}-\right]_y \left[-CH_2-\underset{\underset{CN}{|}}{CH}-\right]_z$$

PBAN

Figure 1.4 Chemical structure of PBAN.

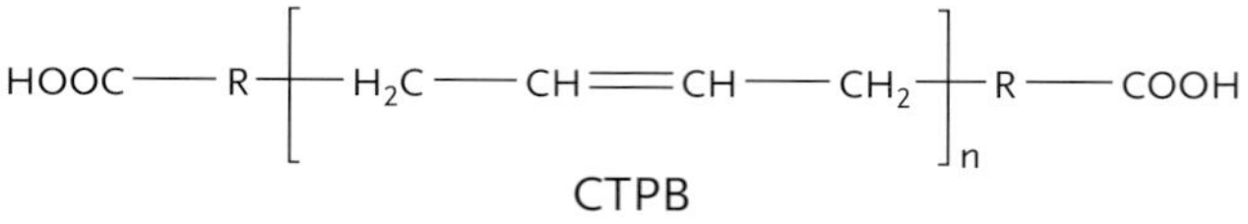

$$HOOC-R-\left[-H_2C-CH=CH-CH_2-R-\right]_n COOH$$

CTPB

Figure 1.5 Chemical structure of CTPB.

used in space transportation system solid rocket motors, with total production exceeding that of all other binder compositions [10].

In late 1950s, carboxyl terminated polybutadiene (CTPB) (Figure 1.5) with the trade name HC-434, which took full advantage of the entire length of the polymer chain, was developed by Thiokol [11].

HC-434 was prepared by the free-radical polymerization using azo-bis-cyanopentanoic acid initiator. In parallel to the Thiokol polymer work, Phillips Petroleum Company developed another brand of CTPB known as Butarez CTL, which was prepared by lithium initiated anionic polymerization. CTPB propellants offer significantly better mechanical properties particularly at lower temperatures in preference to PBAA or PBAN binders, without affecting the specific impulse, density, or solids loading [12]. The curatives for CTPB prepolymers are the same as that for PBAA and PBAN. CTPB formulations were used in propellants from the 1960s. In 1966, the CTPB based propellant TP-H-3062 was used in the surveyor retro motor for the first landing on the moon [13].

1.4
Polyurethanes

Almost concurrently with the development of polybutadiene polymers, Aerojet, who were a competitor to Thiokol, developed the polyurethane branch of binders, in the mid-1950s. Polyurethane binder, the general structure of which is shown in Figure 1.6,

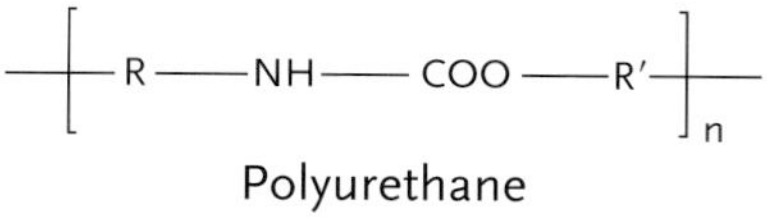

$$\left[-R-NH-COO-R'-\right]_n$$

Polyurethane

Figure 1.6 Chemical structure of polyurethane.

is formed by the reaction of a high molecular weight di-functional glycol with a diisocyanate forming a urethane linked polymer.

The third building block in a polyurethane binder is a triol Such as Trimethylol Propane (TMP) causing cross-linking of polymer chains. Polyurethane binder systems provide shrink-free, low-temperature, and clean cure. An additional benefit of polyurethane binders is that the backbone polymer contains substantial amounts of oxygen [14]. It is not necessary, therefore, to use a high percentage of oxidizer in the formulation of the propellant to achieve comparable energies. Also, several of the urethane polymers are known for their thermal stability [6]. A commonly used polyurethane binder material is ESTANE, a product of the B.F. Goodrich Chemical Company. However, in spite of the advantages of polyurethane binders, polybutadiene formulations still remain more popular.

1.5
Hydroxy Terminated Polybutadiene

The applicability of hydroxy terminated polybutadiene (HTPB) polymer as a binder was demonstrated by Karl Klager of Aerojet in 1961 [4]. HTPB (Figure 1.7) was prepared by the free radical polymerization of butadiene using hydrogen peroxide as the initiator.

Even though the development of HTPB began in 1961, it was not proposed to NASA until 1969 due to the popularity of PBAN and CTPB formulations. HTPB binder was first tested in a rocket motor only in 1972 [4]. It was commercialized with the trade name R-45M by ARCO chemicals. Isocyanates are used as cross-linking agents for HTPB polymers to form urethane linkages, thereby reuniting the polyurethane family of binders with the polybutadiene family. HTPB binders exhibit superior elongation capacity at low temperature and better ageing properties over CTPB [15]. It has since become the most widely-used binder in solid propellant formulations with excellent mechanical properties and enhanced insensitive munition (IM) characteristics [10].

1.6
Explosive Binders

Concurrent with the research on polymers as binders in solid propellants, they were also explored as binders for high explosives, mainly intended for

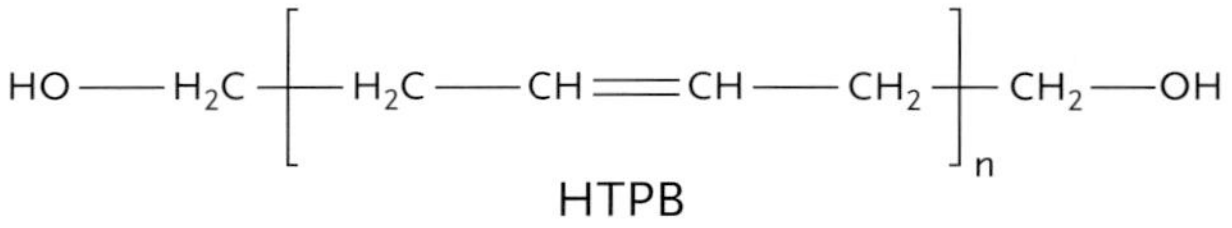

$$HO-H_2C-[H_2C-CH=CH-CH_2]_n-CH_2-OH$$

HTPB

Figure 1.7 Chemical structure of HTPB.

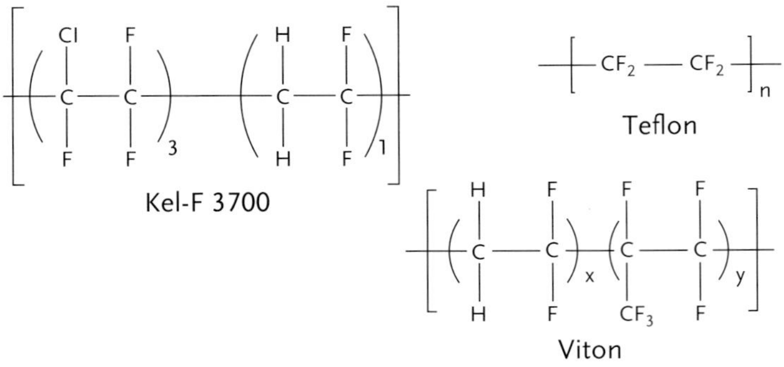

Figure 1.8 Chemical structures of fluoropolymers used as explosive binders.

desensitization. At that time, natural and synthetic waxes were used for desensitization of explosives [16–18]. A more successful method for desensitization was devised through the use of polymer bonded explosives (PBX), in which, the explosive crystals are embedded in a rubber like polymer matrix. The first PBX composition was developed at the Los Alamos Scientific Laboratory, USA, in 1952 [10]. The composition, designated as PBX 9205, consisted of Royal Demolition Explosive (RDX) crystals embedded in a polystyrene matrix plasticized with dioctyl phthalate. It has been found that PBX formulations offer processability and insensitivity advantages over the standard "waxed" explosives [19].

The rubber like polymer binders originally developed for solid propellant applications were successfully applied for PBX formulations also.

With the development of more insensitive explosive fillers (e.g., triaminotrinitrobenzene, TATB), the function of the polymer binder shifted from that of a desensitizer to the one that imparts structural integrity to the PBX formulation. In this context, soft rubbery binders were replaced by the use of hard and high modulus polymers as binders. Epoxy resins, phenolic resins, fluoropolymers (Teflon, Kel-F 3700) (Figure 1.8), and polyamides (Nylon) are examples of hard polymers that are used in PBX formulations.

1.7
Thermoplastic Elastomers

In the early 1980s, screw extrusion technology was envisaged for processing of energetic material formulations, particularly PBX, with the core objective of reducing the cost of production [20]. For its effective implementation, thermoplastic elastomers (TPE) began to be used as binders. TPEs consist of alternate hard and soft segments of crystalline and amorphous polymers, possessing the combined properties of glassy or semi-crystalline thermoplastics and soft elastomers. TPE technology enabled rubbers to be processed as thermoplastics. This feature makes TPEs suitable for high-throughput thermoplastic processes, such as screw extrusion and injection molding, which allow the development and production of

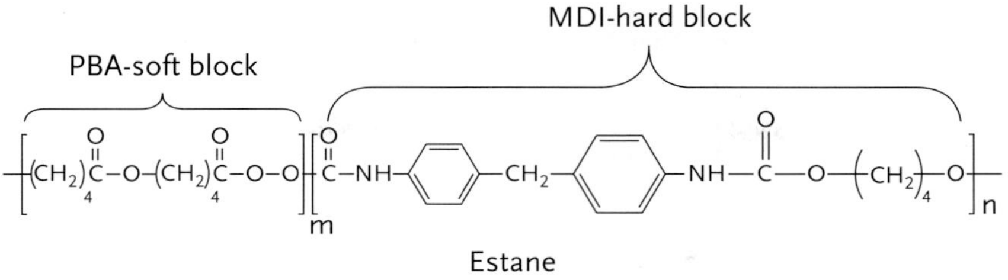

PBA-soft block · MDI-hard block

Figure 1.9 Chemical structure of Estane 5703.

energetic material composites without solvent emissions. Furthermore, TPE binders permit recovery and recycling of energetic material ingredients resulting in additional pollution prevention. TPE binders based on segmented polyurethanes (Estane 5703) and block copolymers of styrene and ethylene/butylene (Kraton G-6500) are widely used as binders in a variety of energetic material formulations including rocket propellants, explosives, and pyrotechnics.

Estane 5703 (Figure 1.9) is a multiblock copolymer obtained by the polymerization of 4,4′-diphenylmethane diisocyanate (MDI) and poly(butylene adipate) (PBA) with 1,4-butanediol as the chain extender. The hard and soft segments of Estane 5703 are polyurethane (MDI) and polyester (PBA), respectively.

Kraton G-6500 (Figure 1.10) is a triblock copolymer of styrene–ethylene/butylene–styrene (SEBS) prepared by anionic polymerization using alkyl lithium initiators.

The polystyrene block is the hard segment and the polyethylene/butylene block constitutes the soft segment. At room temperature, the flexible rubbery polyethylene/butylene blocks ($T_g \sim -100\,°C$) are anchored on both sides by the glassy polystyrene blocks ($T_g \sim 100\,°C$). Therefore, they behave as cross-linked rubber at ambient temperature and allow thermoplastic processing at higher temperatures.

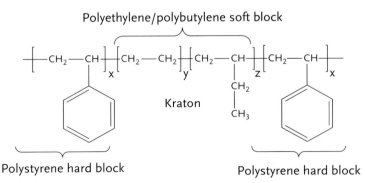

Figure 1.10 Chemical structure of Kraton.

1.8
Energetic Polymers (Other Than NC) as Binders

Strength and sensitivity problems of propellants and explosive formulations were addressed to a larger extent by the success of composite propellant and PBX technologies. However, with the addition of non-energetic or inert binders into formulations, a high level of energetic solid loading is required to meet the given performance requirements, as the explosive energy is diluted. Furthermore, processing technology would also have to be altered in order to cast these highly filled compositions into the required shapes [21].

Hence, in the 1950s, scientists realized the need to develop energetic binders derived from energetic polymers for energetic material formulations. Energetic polymers are obtained through the substitution of energetic functional groups, such as azido and nitrato moieties, as pendent groups to the polymer backbone. The presence of energetic functional groups on the polymer allows the composition to have comparatively less explosive filler, thereby rendering the formulation less sensitive to external stimuli. It is also possible to obtain enhanced performance by using energetic binders instead of the inert ones.

The immediate choice of an energetic polymer for binder application was NC. However, NC suffers from undesirable mechanical properties, particularly very low elongation at sub-ambient temperatures. In order to improve the mechanical/energetic properties of NC formulations, two routes were employed. These were: (i) inert binders in combination with energetic plasticizers, nitroglycerin (NG), or butanetriol trinitrate (BTTN); and (ii) NC binder employed with energetic/non-energetic plasticizers [22]. However, neither of these routes led to formulations with acceptable performance. Therefore, a new series of polyether-based energetic polymers were developed targeted specifically at binder applications.

1.8.1
Polyglycidyl Nitrate

Polyglycidyl nitrate (PGN) (Figure 1.11) was the first energetic prepolymer to be investigated for binder applications. Initial work was done on PGN by Thelen and coworkers [23] in the 1950s at the Naval Surface Warfare Center (NSWC). This was later evaluated as a propellant at the Jet Propulsion Laboratory (JPL) [24].

Development of PGN into an energetic binder was delayed due to the hazardous processes of monomer preparation, purification, and polymerization. The monomer, glycidyl nitrate (GN), was prepared by a single-step method consisting of

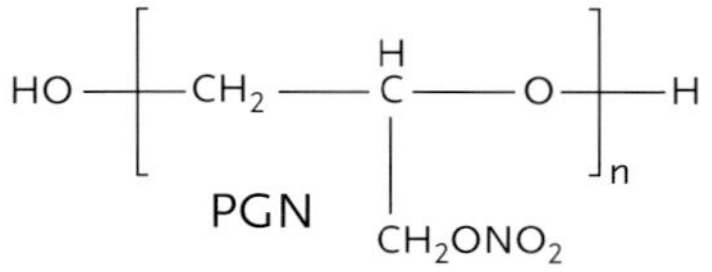

Figure 1.11 Chemical structure of PGN.

reacting glycidol with a potentially dangerous nitrating mixture of 100% nitric acid and acetic anhydride, which is known to generate the unstable explosive acetyl nitrate in-situ. A further disadvantage of the method was the cumbersome purification process of the monomer to remove the dinitroacetate contaminants. Polymerization of GN to PGN was carried out by using entire monomer in the reaction, but this was considered too hazardous due to the exothermic nature of the process.

In the 1990s, the British Defense Research Agency (DRA) modified the monomer preparation by using dinitrogen pentoxide (N_2O_5) in a flow reactor to give dichloromethane solutions of GN in high yield and purity [25, 26]. This process does not require any further purification prior to polymerization. After establishing the method of monomer synthesis, di-functional PGN was safely and reproducibly prepared by the cationic ring opening polymerization of GN. This polymerization employed a tetrafluoroboric acid etherate initiator combined with a di-functional alcohol. The hydroxyl terminated polymers were subsequently cross-linked with isocyanate curing agents to give energetic polyurethanes with potential application as binders in explosives and propellants. Concurrently with the work in the UK on PGN, considerable success was achieved on the scale-up of PGN production and its evaluation as a propellant binder at the Naval Weapons Center, China Lake, USA [27].

1.8.2
GAP

In 1976, research work was initiated at Rocketdyne in the USA on the preparation of a hydroxy-terminated azido prepolymer (glycidyl azide prepolymer, GAP) as an energetic polymer (Figure 1.12), which takes advantage of the positive heat of formation of the azido chemical groups [28].

The logical starting point for GAP synthesis was glycidyl azide (GA), which was prepared by the reaction of epichlorohydrin (ECH) with hydrazoic acid, followed by cyclization with a base. However, attempts to polymerize GA were unsuccessful due to the lack of reactivity of the monomer. Emphasis shifted to polymerization of ECH to give polyepichlorohydrin (PECH), followed by conversion of PECH into GAP. GAP triol was successfully prepared in 1976 by the reaction of PECH triol with sodium azide in a dimethylformamide medium [28]. GAP is a unique high-density polymer with a positive heat of formation equal to +490.7 kJ/mol. Currently, GAP is the most readily available energetic polymer due to the low cost of

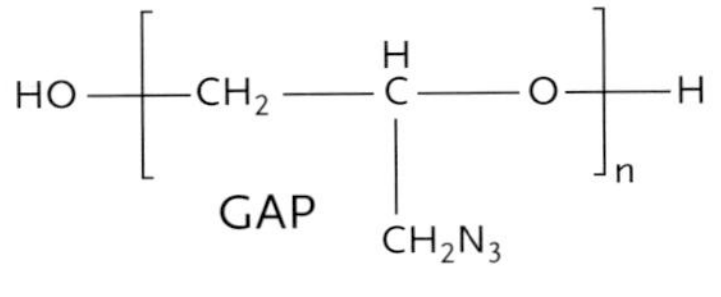

Figure 1.12 Chemical structure of GAP.

the synthetic route and its excellent binder properties [29]. The 3M Company has commercialized the GAP polymer under the name GAP 5527 polyol.

1.8.3 Energetic Polyoxetanes

Energetic polymers derived from oxetane monomers, namely 3,3-bis(azidomethyl) oxetane (BAMO), 3-azidomethyl 3-methyl oxetane (AMMO), and 3-nitratomethyl methyl oxetane (NIMMO), were sought for binder applications, because of their low viscosity and good mechanical properties after cross-linking. G.E. Manser discovered energetic polyoxetanes based on BAMO and AMMO at Aerojet in 1984 [30]. His group subsequently reported the preparation and polymerization of the nitrato alkyl oxetane monomer, NIMMO [31, 32]. The energetic polyoxetanes (Figure 1.13) were synthesized by the cationic ring opening polymerization of the respective monomers using borontrifluoride etherate catalyst [33–35].

The critical aspect of the preparation of energetic polyoxetanes is the ease of preparation and purity of monomers. NIMMO was first prepared by the nitration of 3-hydroxy methyl-3-methyl oxetane (HMMO) by acetyl nitrate [33]. Owing to the hazardous nature of the reaction, the synthesis was modified by selective nitration of HMMO using dinitrogen pentoxide nitrating agent in a flow nitration system at DRA [36], which provided excellent yields of pure NIMMO. Therefore, among the energetic polyoxetanes, Poly(NIMMO) gained popularity due to its scalable and safe procedure for preparation [37]. Poly(NIMMO) is a very promising binder for propellant and explosive applications. The manufacturing process for Poly (NIMMO) is licensed to ICI, UK, by the DRA.

Synthesis of BAMO monomer involved treating 3,3-bis(chloromethyl) oxetane (BCMO) with sodium azide in dimethylformamide [38]. The monomer AMMO

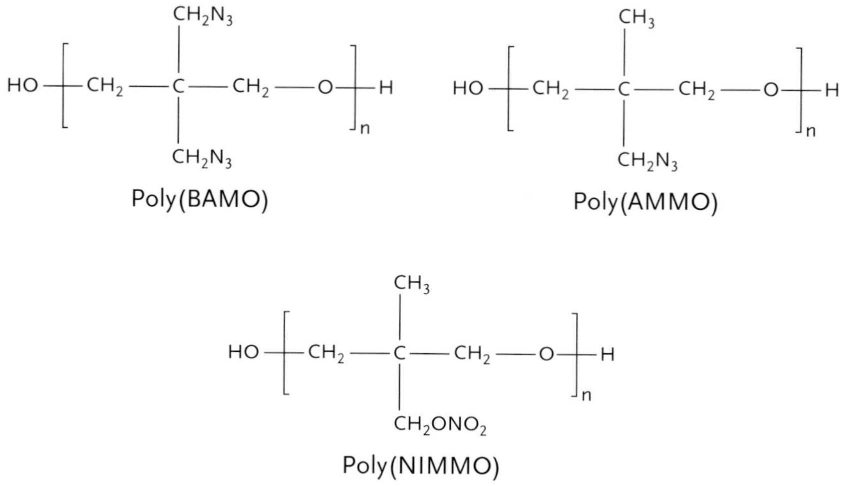

Figure 1.13 Chemical structures of polyoxetanes used as energetic binders.

was prepared by azidation with sodium azide of the tosylate derivative of HMMO [39]. The symmetrically di- substituted Poly(BAMO) is too highly crystalline to be used as a homopolymer for binder applications. Hence it must be co-polymerized with the relatively less energetic AMMO or NIMMO to bring down the melting and glass transition temperatures. Poly(BAMO) is nearing commercialized production by Aerojet and Thiokol.

Energetic polyoxetanes containing difluoroamine ($-NF_2$) groups were successfully synthesized on the laboratory scale by the cationic ring opening polymerization of 3,3-bis(difluoroaminomethyl) oxetane or 3-difluoroaminomethyl 3-methyl oxetane using borontrifluoride etherate catalyst [40]. However, the difficult synthetic steps in the monomer preparation have so far prevented their evaluation as binders in large-scale.

1.8.4
Polyphosphazenes

Recently, inorganic polymers based on polyphosphazenes have shown promise as energetic binders on account of their high densities, low glass transition temperatures, potential synthetic flexibilities, and good chemical and thermal stabilities. Polyphosphazenes, having the general structure $(N=PR_2)_n$, (Figure 1.14) are inorganic–organic polymers in which the side groups (R) can be halogeno, or organo units [41].

Phosphazene polymers are rendered energetic by the macromolecular replacement of halogen/organo units by nitrato or azido groups. The synthetic pathway comprises the use of polymeric alkoxy substituted precursors of phosphazenes and its subsequent nucleophilic substitution of the alkoxy group with the energetic pendant group. Both nitrate ester and azide functionalized energetic phosphazenes have been successfully synthesized on a laboratory scale at the Atomic Weapons Establishment (AWE), UK, for potential binder applications [42].

1.8.5
Energetic Thermoplastic Elastomers

Energetic versions of TPE binders (energetic thermoplastic elastomers, ETPEs) have been developed by Thiokol Inc. USA, to be used as binders in melt cast explosive and propellant formulations [43]. ETPEs consist of alternate crystalline (hard) and amorphous (soft) segments of energetic polymer molecules. The hard

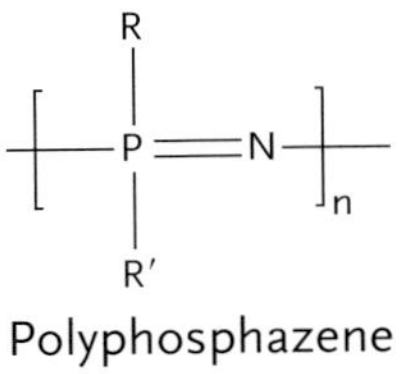

Polyphosphazene

Figure 1.14 General chemical structure of polyphosphazene.

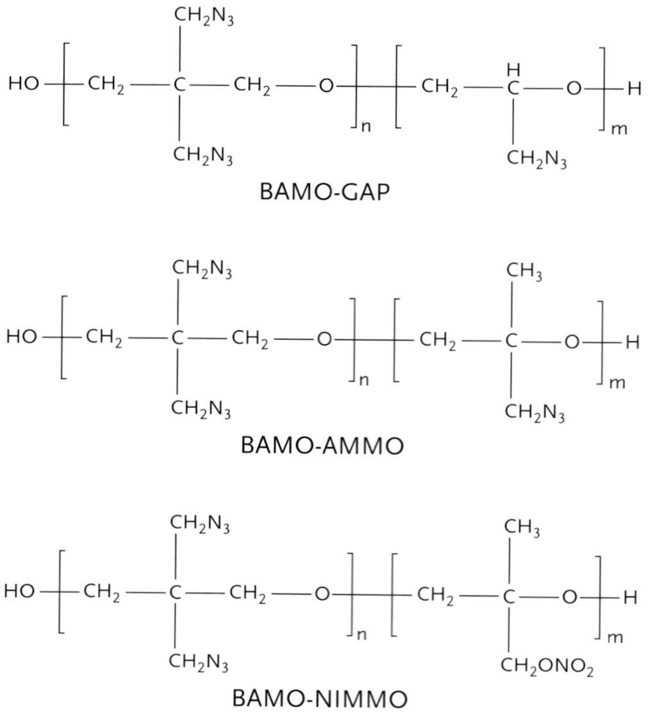

Figure 1.15 Chemical structures of various ETPEs used as energetic binders.

segment of ETPEs consist of Poly(BAMO) and the soft segment of Poly(NIMMO), Poly(AMMO) or GAP. Typical examples of ETPEs are illustrated in Figure 1.15.

ETPEs are prepared by either linking the blocks of individual energetic polymers with isocyanates [44] or by sequential polymerization [45], and have been used as binders in experimental formulations of new low-vulnerability (LOVA) propellants with success [46, 47]. They are environmentally friendly and recyclable. The utilization of ETPE as a binder is rapidly increasing with the emergence of twin-screw extrusion as a promising route for the manufacture of ETPE based formulations.

1.9
Energetic Polymer Plasticizers

Generally, plasticizers are non-reactive liquid diluents used for improving the processability and low temperature mechanical properties of energetic material composites. Energetic plasticizers based on nitratoesters (e.g., BTTN) not only improve processing and low-temperature properties but also improve the overall energetic properties of the formulation. However, nitrate ester plasticizers suffer from migration problems, especially with energetic binder formulations, resulting in the loss of the plasticizer over a period of time. A promising recent approach is to use

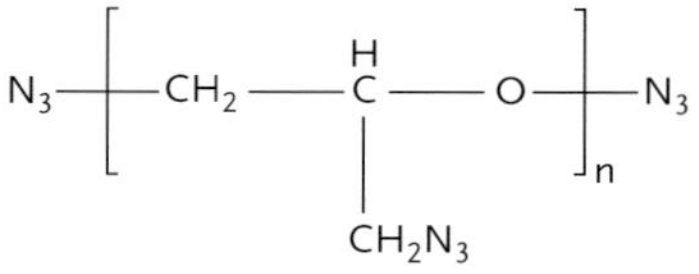

$$N_3 \!-\!\!\left[\!-\!CH_2 \!-\!\!\underset{CH_2N_3}{\overset{H}{\underset{|}{C}}} \!-\!O \!-\!\right]_{\!n}\!\!-\! N_3$$

Figure 1.16 Chemical structure of azido terminated GAP.

low molecular weight oligomers of energetic polymers for plasticizer applications, which offer a number of advantages, including excellent miscibility with the new energetic binders, low volatility, low glass transition temperature, decreased plasticizer mobility, excellent combustion properties, and reduced hazard characteristics [48, 49]. Under this category, low molecular weight GAP [50] and GLYN (glycidyl nitrate) polymers [51] have gained importance as energetic polymer plasticizers.

In another important development [52], the molecular structure of the low molecular weight GAP polymer have been modified to convert the free hydroxyl moieties at the chain ends into azido functional groups (Figure 1.16).

This will prevent unwanted reaction of the plasticizer with the isocyanate cross-linking agent, which results in the loss of plasticizing action. The 3M Company has commercialized this product as GAP-0700 plasticizer.

The timeline of the development of polymers for binder applications is illustrated in Scheme 1.1 and the details are given in Table 1.1.

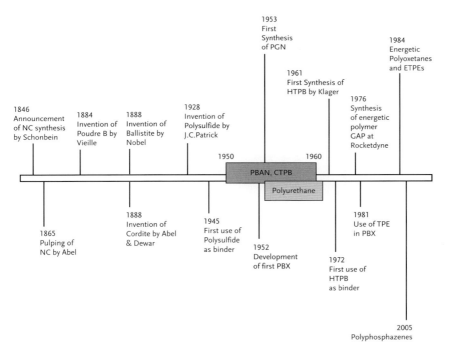

Scheme 1.1 Time line of the development of different polymers as binder for propellant and explosive applications.

Table 1.1 Polymers used in binder applications.

Polymer (abbreviation)	Preparation method	Trade name	Curing agent	Year of development
Nitrocellulose (NC)	Nitration of cellulosic materials (cotton, paper, wood)	Poudre B; Ballistite; Cordite	—	1888
Polysulfide elastomer	Condensation polymerization of ethylene dichloride and sodium polysulfide	Thiokol LP series	p-Quinonedioxime	1928
Polybutadiene–acrylic acid copolymer (PBAA)	Emulsion radical polymerization	HA series	Di- or tri-functional imine/epoxide	1954
Polybutadiene–acrylic acid–acrylonitrile copolymer (PBAN)	Emulsion radical polymerization	HB series	Di- or tri-functional imine/epoxide	1957
Carboxyl terminated polybutadiene (CTPB)	Free radical polymerization	HC-434; Butarez CTL	Di- or tri-functional imine/epoxide	Late 1950s
Polyurethane (PU)	Condensation polymerization of diols and isocyanates	Estane	Multifunctional alcohols	Mid-1950s
Hydroxyl terminated polybutadiene (HTPB)	Free radical polymerization of butadiene	R-45M	Isocyanates	1961
Energetic polyglycidyl nitrate	Ring opening polymerization of glycidyl nitrate	PGN	Isocyanates	1950 Commercialized by DRA in 1990
Glycidyl azide polymer	Ring opening polymerization of epichlorohydrin followed by azidation	GAP 5527	Isocyanates	1971
Energetic polyoxetanes	Ring opening polymerization of oxetanes	Poly(oxetanes)	Isocyanates	1990

References

1. Davis, T.L. (1972) *The Chemistry of Powder and Explosives*, Angriff Press, p. 548.
2. Perrson, P.-A., Holmberg, R., and Lee, J. (1993) *Rock Blasting and Explosives Engineering*, CRC Press, p. 67.
3. Davenas, A. (2003) Development of modern solid propellants. *J. Propul. Power*, **19** (6), 1108.
4. Hunley, J.D. (1999) History of solid propellant rocketry. What we do and do not know. 35th AIAA/SAE/ASME Joint Propulsion Conference and Exhibit, Los Angeles, California. June 20–24.
5. Sutton, E.S. (1999) From polymers to propellants to rockets: A history of Thiokol. AIAA/SAE/ASME Joint Propulsion Conference and Exhibit, Los Angeles, California. June 20–24.
6. Arendale, W.F. (1967) Chemistry of propellants based on chemically crosslinked binders. Propellants manufacture, hazards and testing, in *Advances in Chemistry Series*, Vol. **88** (eds C. Boyars and K. Klager), American Chemical Society, Washington D.C., pp. 67–83.
7. Mastrolia, E.J. and Klager, K. (1967) Solid propellants based on polybutadiene binders. Propellants manufacture, hazards and testing, in *Advances in Chemistry Series*, Vol. **88** (eds C. Boyars and K. Klager), American Chemical Society, Washington D.C., pp. 122–164.
8. Hendel, F.J. (1965) Review of solid propellants for space exploration. NASA Technical Memorandum No: 33–254, Jet Propulsion Laboratory, Pasadena, California.
9. Cesaroni, A.J. (2004) Thermoplastic polymer propellant compositions. US Patent 6,740,180.
10. Daniel, M.A. (2006) Polyurethane binders for polymer bonded explosives. DSTO-GD-0492.
11. Sutton, E.S. (1984) From polysulfides to CTPB binders – A major transition in solid propellant binder chemistry. AIAA/SAE/ASME Joint Propulsion Conference, Cincinnati, Ohio.
12. Chaille, J.L. Development of a composite rocket propellant. Technical Report No: S-64, Rohm and Haas Company, Huntsville, AL, USA.
13. Moore, T. and Rohrbaugh, E. (2002) CTPB propellants for space applications. 38th AIAA/ASME/SAE/ASEE. Joint Propulsion Conference, Indianapolis, Indiana.
14. Klager, K. (1984) Polyurethanes the most versatile binder for solid composite propellants. AIAA/SAE/ASME Joint Propulsion Conference, Cincinnati, Ohio.
15. Moore, T.L. (1997) Assessment of HTPB and PBAN propellant usage in the United States. 33rd AIAA/ASME/SAE/ASEE. Joint Propulsion Conference, Seattle, Washington.
16. Copp, J.L. and Ubbelohde, A.R. (1948) The grit sensitiveness of high explosives. *Philos. Trans. R. Soc. London. Ser. A. Math. Phy. Sci.*, **241** (831), 248–265.
17. Linder, P.W. (1961) Desensitization of explosives. *Trans. Faraday Soc.*, **57**, 1024–1030.
18. Kosowski, B.M. and Taylor, R.C. (1995) Method and composition for melt cast explosives, propellants and pyrotechnics. US Patent 5,431,756.
19. Kneisl, P. (2006) Hi-temp explosive binder. US Patent 6,989,064.
20. Fong, C.W. (1986) Manufacture of propellants and polymer bonded explosives by twin screw extrusion, Technical Report: 0456, Weapons Systems Research Laboratory.
21. Colclough, M.E., Desai, H., Millar, R.W., Paul, N.C., Stewart, N.J., and Golding, P. (1993) Energetic polymers as binders in propellants and explosives. *Polym. Adv. Technol.*, **5**, 554.
22. Antic, G. and Dzingalasevic, V. (2006) Characteristics of cast PBX with aluminum. *Sci. Tech. Rev.*, **LVI** (3–4), 52–58.
23. (a) Murbach, W.J., Fish, W.R., and Van Dolah, R.W. (1953) U.S. Naval Ordnance NAVORD Report 2028, polyglycidyl nitrate. Part 1 Preparation and characterization of glycidyl nitrate.

NOTS, **685**. (b) Meitner, J.G., Thelen, C.J., Murbach, W.J., and Van Dolah, R.W. (1953) US Naval Ordnance NAVORD Report 2028, polyglycidyl nitrate. Part 2 Preparation and characterization of glycidyl nitrate. NOTS, **686**.

24 Ingham, J.D. and Nichols, P.L. Jr (1959) High performance PGN-polyurethane propellants, Publication Number 93. Jet Propulsion Laboratory.

25 Desai, H.J., Cunliffe, A.V., Lewis, T. Millar, R.W., Paul, N.C., Stewart, M.J., and Amass, A.J. (1996) Synthesis of narrow molecular weight hydroxy telechelic polyglycidyl nitrate and estimation of theoretical heat of explosion. *Polymer*, **37** (15), 3471–3476.

26 Cumming, A. (1997) New directions in energetic materials. *J. Defence Sci.*, **1** (3), 319.

27 Willer, R.L. (2009) Calculation of the density and detonation properties of C, H, N, O and F compounds: use in the design and synthesis of new energetic materials. *J. Mex. Chem. Soc.*, **53** (3), 108–119.

28 Frankel, M.B., Grant, L.R., and Flanagan, J.E. (1989) Historical development of GAP. ASME, SAE and ASEE 25th Joint Propulsion Conference, Monterey, California.

29 Provatas, A. (2000) Energetic polymers and plasticizers for explosive formulations – A review of recent advances. DSTO-TR-0966.

30 Manser, G.E. (1984) Energetic copolymers and methods of making the same. US Patent 4,483,978.

31 Manser, G.E. (1995) Latest advancements in binder development. Proceedings of JANNAF Novel Ingredients for Liquid and Solid Propellants. Specialist Session. CPIA, PDCS-SS-01, 16–38.

32 Manser, G.E. and Hajik, R.M. (1993) Method of synthesizing nitrato alkyl oxetanes. US Patent 5,214,166.

33 Earl, R.A. and Elmslie, J.S. (1983) Preparation of hydroxy-terminated poly (3,3-bisazidomethyloxetanes). US Patent 4,405,762.

34 Desai, H. (1996) Telechelic polyoxetanes. The polymeric materials encyclopedia: synthesis, *Properties and Applications*, vol **11** (ed. J.C. Salamone), CRC Press, pp. 8268–8279.

35 Manser, G.E. and Ross, D.L. (1982) Synthesis of energetic polymers. ONR final report. Aerojet Solid Propulsion Company USA.

36 Miller, R.W., Colclough, M.E., Golding, P., Honey, P.J., Paul, N.C., Sanderson, A.J., and Stewart, M.J. (1992) New synthesis routes for energetic materials using dinitrogen pentoxide. *Philos. Trans.: Phys. Sci. Eng.*, **339** (1654), 305–319.

37 Agrawal, J.P. (2005) Some new high energy materials and their formulations for specialized applications. *Propellants Explos. Pyrotech.*, **30** (5), 315.

38 Frankel, M.B. and Wilson, E.R. (1981) Energetic azido monomers. *J. Chem. Eng. Data*, **26**, 219.

39 Barbieri, U., Polacco, G., Paesano, E., and Massimi, R. (2006) Low risk synthesis of energetic poly (AMMO) from tosylated precursors. *Propellants Explos. Pyrotech.*, **31** (5), 369–374.

40 Manser, G.E. and Archibald, T.A. (1993) Difluoramino oxetanes and polymers formed from there for use in energetic formulations. US Patent 5,272,249.

41 Allcock, H.R., Reeves, S.D., Nelson, J. M., and Manners, I. (2000) Synthesis and characterization of phosphazene di- and triblock copolymers via the controlled cationic, ambient temperature polymerization of phosphoranimines. *Macromolecules*, **33**, 3999–4007.

42 Golding, P., Trussel, S.J., and Beckham, R.W. (2006) Novel energetic polyphosphazenes. WO 2006/0322882 A1.

43 Manser, G.E. and Miller, S.R. (1992) Thermoplastic elastomers having alternate crystalline structure for use as energetic binders. US Patent 5,210,153.

44 Wardle, R.B. (1989) Method of producing thermoplastic elastomers having alternate crystalline structure for the use as binders in high-energy compositions. US Patent 4,806,613.

45 Wardle, R.B., Cannizo, L.F., Hamilton, R.S., and Hinshaw, J.C. (1992) Energetic oxetane thermoplastic elastomer

binders. ONR Technical Report No: AD-A278 307.

46 Beaupr'eet, F., Ampleman, G., Nicole, C., and Melancon, J.G. (2003) Insensitive propellant formulations containing energetic thermoplastic elastomers. US Patent 6,508,894.

47 Horst, A.W., Johnson, L.D., May, I.W., and Morrison, W.F. (1997) Recent advances in anti-armor technology. AIAA Paper 97–0484, AIAA, 35th Aerospace Science Meeting and Exhibit, Reno, Nevada, USA.

48 Willer, R.W., Stearn, A.G., and Day, R.S. (1995) PGN plasticizers. US Patent 5,380,777.

49 Bala, K. and Golding, P. (2004) Influence of molecular weight on explosive hazard. Proceedings of NDIA. Insensitive Munitions and Energetic Materials Symposium.

50 Ahad, E. (1990) Direct conversion of epichlorohydrin to glycidyl azide polymer. US Patent 4,891,438.

51 Provatas, A. (2003) Characterization and binder studies of the energetic plasticiser–GLYN oligomer. DSTO-TR-1422.

52 Ampleman, G. (1992) Synthesis of diazido terminated energetic plasticizer. US Patent 5,124,463.

2
High Nitrogen Content Polymers

2.1
Introduction

High nitrogen content compounds are a unique class of novel energetic materials deriving most of their energy from the high positive heat of formation rather than from the oxidation of the carbon backbone [1]. Energetic polymers with high nitrogen content are obtained by substituting the energetic azido and tetrazole groups in the side chains of macromolecules. The most important characteristic of the high nitrogen content polymers is their positive heat of formation resulting in exothermicity during binder degradation. Furthermore, the high nitrogen content of the polymer enhances the density and produces large amounts of fuel rich hot gases per mole of polymer during thermal decomposition. This chapter discusses the preparation, curing, thermal decomposition behavior, and combustion characteristics of azido polymers finding potential applications as energetic binders. Comparisons are provided *vis-à-vis* the properties of hydroxy terminated polybutadiene (HTPB) – the widely used prepolymer for inert binder formulations. Energetic tetrazole polymers are also briefly discussed at the end of the chapter.

2.2
Preparation of Energetic Azido Polymers

2.2.1
Glycidyl Azide Polymer

Linear hydroxy terminated GAP prepolymers have been successfully synthesized in a two-step process [2–4]. The first step is the synthesis of polyepichlorohydrin (PECH) diol of adequate molecular weight. In this step, epichlorohydrin (ECH) is polymerized to polyepichlorohydrin (PECH) by cationic ring opening polymerization in the presence of a Lewis acid catalyst (e.g., boron trifluoride etherate) in dichloromethane. The second step involves the azidation of PECH using a molar excess of sodium azide (NaN_3) in the presence of an organic aprotic (e.g., dimethyl sulfoxide, DMSO) or aqueous [5] solvent (Scheme 2.1). The aqueous

Energetic Polymers: Binders and Plasticizers for Enhancing Performance, First Edition.
How Ghee Ang and Sreekumar Pisharath.
© 2012 WILEY-VCH Verlag GmbH & Co. KGaA, Weinheim. Published 2012 by WILEY-VCH Verlag GmbH & Co. KGaA

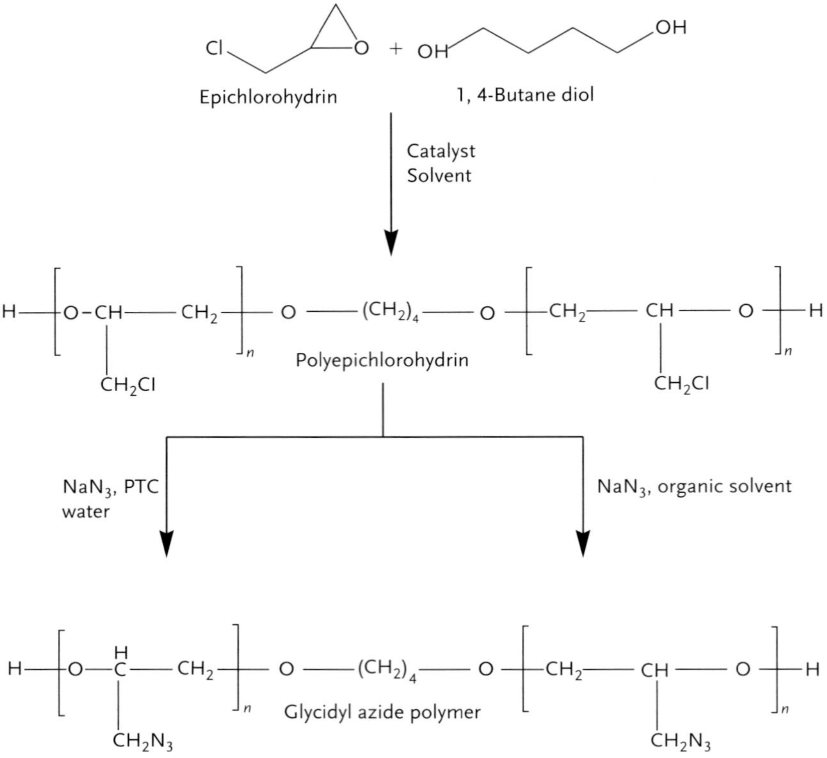

Scheme 2.1 Synthetic scheme for GAP.

process consists of a phase transfer catalyst (PTC): methyl tricapryl ammonium chloride. The polymer is obtained as a yellow viscous liquid with an average hydroxyl functionality of two. The nature of the hydroxyl groups is mainly secondary. A significant slowing down of the reaction has been observed after 90% conversion, as a consequence of the association of the metal cation of the azide salt with the solvent medium.

A solvent free molten salt method was developed by Wagner [6] for the synthesis of GAP, which uses low melting quaternary ammonium azide salts instead of ionic azides (Scheme 2.2). The reaction rate for the PECH to GAP conversion is directly proportional to the concentrations of epichlorohydrin (ECH) units in the polymer and azide ion. GAP at a yield of more than 90% could be obtained in 3 h by the molten salt method. However, in large-scale preparations, the safe operation of the reaction is problematic, as the sequential decomposition temperature of the quaternary ammonium azide salts is closer to the decomposition temperature of GAP. Therefore, GAP is still commercially prepared by the aprotic solvent process.

The curing reaction of GAP diol with isocyanate curing agents is dependent on the nature of the end hydroxyl groups and the number of such hydroxyl groups in

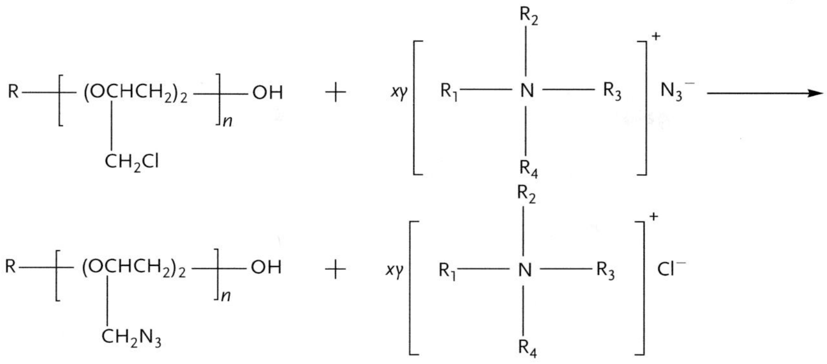

Scheme 2.2 Schematic of the molten salt method for GAP synthesis.

the polymer chain. Primary hydroxyl groups are preferred over the secondary, as they provide faster curing reactions at lower temperatures and also eliminate the gassing problems that exist with the secondary hydroxyl groups. The reactivity of the terminal secondary hydroxyl groups in linear glycidyl azide polymers is equal to the reactivity of water towards isocyanate. Therefore, water can react with the isocyanate in the curing reaction causing gas evolution, which results in cracks and bubbles in the cured propellants. Also, it is highly desirable to have GAP with a higher functionality (more than two) to avoid the use of extra cross-linking agents, such as triols and triisocyanates.

Ampleman [7] patented a process to synthesize GAP bearing primary hydroxyl groups and higher functionality, which involves regiospecific epoxidation of linear PECH under basic conditions. Subsequent opening of the epoxides under different conditions leads to the formation of PECH polymers with various functionalities (Scheme 2.3). As shown in the scheme, ring opening in the presence of acidic water doubles the hydroxyl value of PECH. Similarly, ring opening with trimethylol propane triples the functionality, and with pentaerythritol quadruples the functionality of polymer. Finally, azidation of these polymers yields GAP polymers with high functionality and reactivity.

The branched hydroxyl terminated GAP binders (Figure 2.1) have higher functionality, lower glass transition temperature, and lower viscosity as compared with the linear GAP diol (Table 2.1). Hence faster curing rates, better mechanical properties, and easy processability could be expected for branched GAP based binders.

Ahad [8] prepared branched hydroxyl terminated GAP in a single step through a simultaneous degradation and azidation procedure. In the process, high molecular weight PECH is reacted with an alkali metal azide in organic solvent and the polymer chain is cleaved with a basic catalyst to form branches. Typical cleaving catalysts are lithium methanolate and sodium ethoxide. The molecular weight of branched GAP is controlled by adjusting the ratio of the cleaving catalyst to PECH rubber. A pilot plant to manufacture 5 kg/batch of branched GAP is in operation at the Defense Research Establishment, Val Cartier, Canada [9, 10].

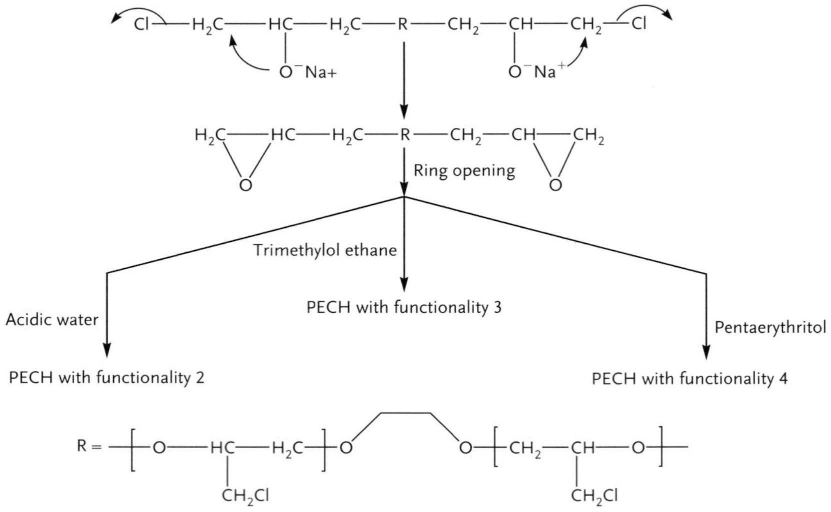

Scheme 2.3 Synthetic scheme to prepare PECH of various functionalities [7].

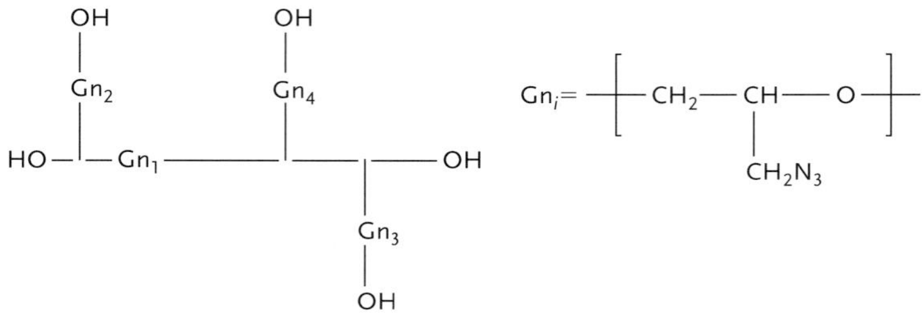

Figure 2.1 General structure of branched GAP polymer.

Table 2.1 Comparative properties of linear GAP and branched GAP [7].

Polymer	Viscosity (cP)	T_g (°C)	Molecular weight	Functionality
Linear GAP	10 000	−50	3000	~2
Branched GAP	4500	−60	4200	>2

2.2.2
Azido Polymers from Oxetanes

Azido polymers derived from oxetane monomers have gained importance as energetic binders because they are expected to offer low viscosity and good mechanical properties on cross-linking [11]. In this section the preparation of two important energetic azido polymers; Poly(BAMO) (3,3-bis(azidomethyl) oxetane) and Poly(AMMO) (3-azidomethyl 3-methyl oxetane) are discussed.

2.2.2.1 Poly(BAMO)

Poly(BAMO) is synthesized by two reaction procedures (Scheme 2.4). In one of the methods, the monomer 3,3-bis(chloromethyl) oxetane (BCMO) is treated with sodium azide in dimethylformamide (DMF) at 90 °C for 2 h. The 3,3-bis(azidomethyl) oxetane (BAMO) thus obtained is polymerized by cationic ring opening polymerization using boron trifluoride (BF_3)-etherate as catalyst in the presence of a butanediol initiator at −5 °C [12, 13]. Polymerization of BAMO is hazardous, as BAMO is an explosive liquid monomer with almost the same shock sensitivity as nitroglycerin [12].

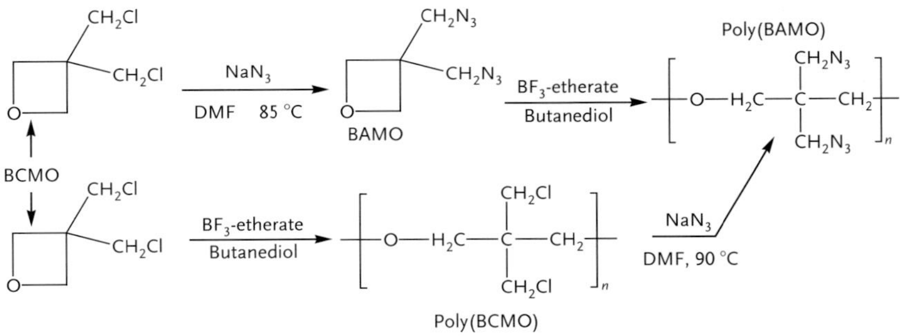

Scheme 2.4 Synthetic schemes for the preparation of Poly(BAMO).

Hence an alternative route for the preparation of Poly(BAMO) was suggested, which involves the ring opening polymerization of BCMO using butanediol as initiator and BF_3-etherate as the catalyst at −5 °C [14, 15]. The obtained polymer is reacted with sodium azide in DMF solvent at 90 °C. BCMO is obtained by the ring closure reaction of the trichloro derivative of pentaerythritol [16].

Poly(BAMO) has a strong tendency to crystallize, which could be detrimental to the mechanical properties of the formulation [17]. Hence Poly(BAMO) is copolymerized with other monomers (tetrahydrofuran, 3-chloromethyl-3-(2,5,8-trioxadecyl) oxetane, caprolactone) to disrupt its chain regularity and also to lower the glass transition temperature [12, 15, 18, 19]. In this way, the energetic properties of Poly(BAMO) have been sacrificed for better processability and mechanical properties of the binder. Poly(BAMO) has been copolymerized with energetic monomers to provide energetic binders with thermoplastic elastomer (TPE) behavior. This topic is dealt in greater detail in Chapter 4 on energetic thermoplastic elastomers.

2.2.2.2 Poly(AMMO)

Poly(AMMO) (3-azidomethyl 3-methyl oxetane) is synthesized by the cationic ring opening polymerization of the AMMO monomer (Scheme 2.5). AMMO is prepared through two synthetic routes starting from 3-hydroxy methyl-3-methyl oxetane (HMMO).

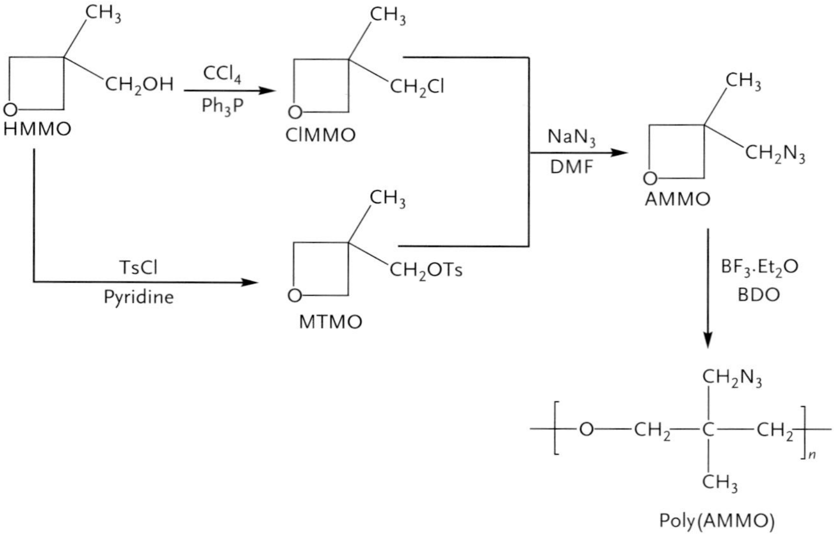

Scheme 2.5 Synthetic scheme for the preparation of Poly(AMMO).

The differences between the preparation schemes are in the intermediates from which the AMMO monomer is prepared. In the first scheme, AMMO is prepared by the azidation of ClMMO using sodium azide and dimethylformamide. ClMMO is synthesized by the chlorination of HMMO using a carbon tetrachloride/triphenylphosphine mixture [20]. Carbon tetrachloride serves as the both solvent and halogen source in the reaction.

The second route involves nucleophilic displacement of the tosylate derivative of HMMO, 3-methyl-3′-(tosyloxymethyl) oxetane (MTMO), by azide to obtain AMMO [21]. MTMO is prepared by the reaction of HMMO with *p*-toluenesulfonyl chloride in the presence of pyridine. AMMO is polymerized by cationic ring opening polymerization using boron trifluoride (BF_3)-etherate as catalyst in the presence of butanediol initiator to obtain Poly(AMMO) [22].

In another effort, a difunctional cationic initiator system, bis(chlorodimethylsilyl) benzene/silver hexafluoroantimonate in methylene chloride at −78 °C, was used to polymerize AMMO [23]. Reaction offered the advantage of a quasi living cationic polymerization, in which the number average molecular weight (M_n) of the polymer was found to increase linearly with monomer to initiator ratio. Consequently, it became possible to produce energetic polymers with predetermined functionality and molecular weight.

2.3
Physical Properties of Azido Polymers

The important physical properties of energetic azido polymers are compared with inert HTPB as a reference in Table 2.2.

All the azido polymers possess higher densities, positive heats of formation, and higher oxygen balances as compared with the inert HTPB. These properties mean that azido polymers are good binder candidates for explosive and propellant formulations. However, the lower glass transition temperature of HTPB provide it with better low temperature mechanical properties. Research efforts are progressing to improve the mechanical properties of azido polymers by the addition of compatible energetic plasticizers or through incorporation of flexible linear structural polymers [24].

Table 2.2 Comparison of properties of azido energetic polymers with the inert HTPB.

Polymer	Hydroxyl functionality	Density (kg/m^3)	T_g (T_m) (°C)	Heat of formation (kJ/kg)	Oxygen balance (%)
GAP	2	1.34	−34	+957	−121.1
Poly(BAMO)	2	1.34	−34 (80)	+2460	−123.2
Poly (AMMO)	2	1.1	−42	+345.3	−170
HTPB	2	0.9	−68	−582	−325

2.4
Curing of Azido Polymers

Curing is the process in which a prepolymer is transformed irreversibly into an infusible and insoluble three dimensional polymer network, which is capable of supporting dynamic and also static stresses. Generally, curing of the azido polymers is accomplished by the reaction of the terminal hydroxyl groups of the prepolymer with diisocyanate or polyisocyanate curing agents to form a polyurethane network. Other than polyisocyanates, a new curing methodology using dipolarophile molecules (substituted alkenes) is also under investigation for azido polymers.

2.4.1
Curing by Polyisocyanates

Energetic prepolymers should be cross-linked to produce a tough elastomeric matrix before they are employed as binders in formulations. Cross-linking density is a critical parameter that determines the mechanical properties, especially the elasticity of the polymer binder. Hydroxyl terminated azido polymers are cross-linked through the formation of polyurethane networks by isocyanate curing agents.

2.4.1.1 Preparation

The two usual routes for preparation of polyurethane networks using isocyanate curing agents are the prepolymer method and the one-shot method. In the prepolymer method, the hydroxyl terminated energetic polymer and the polyisocyanate are reacted in the presence of a catalyst to form an intermediate prepolymer, which is a thick viscous fluid.

The excess of NCO groups at the chain ends of the energetic polyurethane prepolymer are utilized for subsequent chain-extension reactions with low molecular weight polyfunctional alcohols to form high molecular weight polyurethane networks (Scheme 2.6). The common chain-extending reagents used for polyurethanes are 1,4-butanediol, trimethylol propane, glycerol, and pentaerythritol.

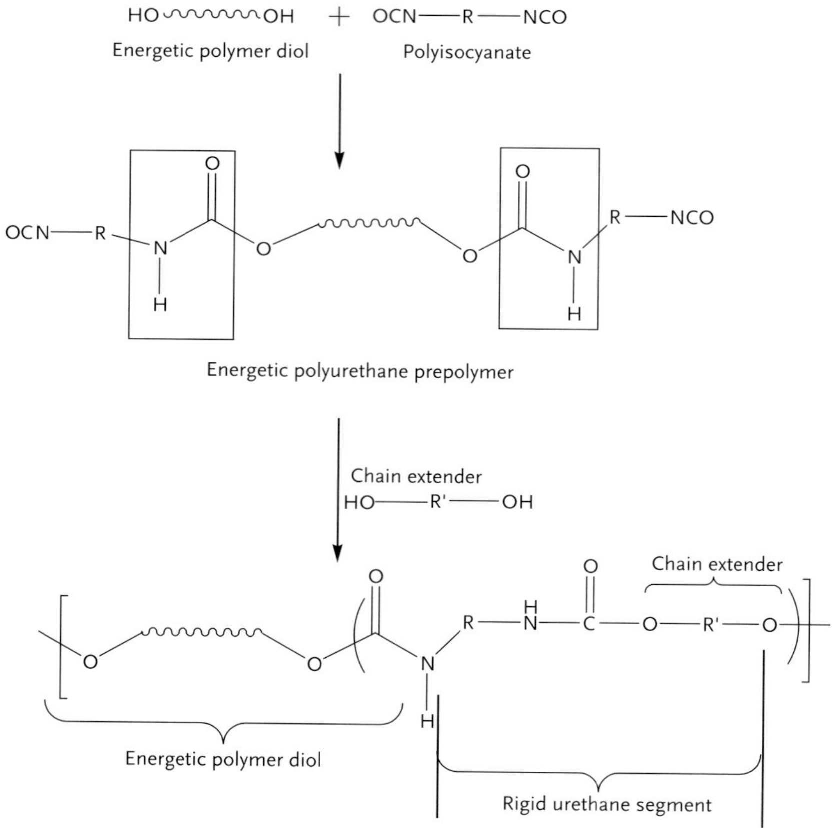

Scheme 2.6 Schematic of curing reaction using prepolymer method.

In the one-shot method, the polyurethane network formation is carried out by the simultaneous reaction of hydroxyl terminated energetic polymer, polyisocyanate, chain extender, and catalyst. There is no formation of the intermediate prepolymer.

In both of the methods, the final product is an energetic cross-linked polyurethane elastomer. Polyurethane elastomers with better structural regularity could be obtained from the prepolymer method resulting in good low temperature properties.

Common polyisocyanates used for curing azido polymers are presented in Figure 2.2.

There is a substantial difference in reactivity between the isocyanate groups in the same molecule as a result of electronic and steric effects, which is important to ensure the completion of the curing reaction. For example, with aliphatic isocyanates, primary isocyanates react faster than secondary isocyanates, consequently in IPDI (isophorone diisocyanate) the primary isocyanate will react first. The ortho substituents on aromatic isocyanates in TDI (toluene 2,4-diisocyanate) are less reactive than an isocyanate in the para 4-position due to steric hindrance. Aromatic isocyanates are more reactive than aliphatic ones due to the electron withdrawing effect of the aromatic ring systems.

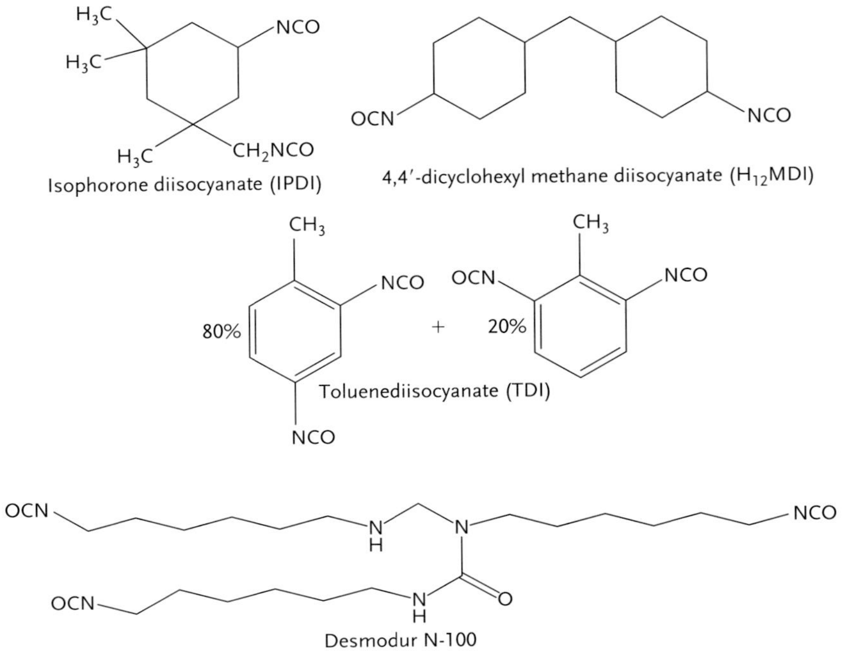

Figure 2.2 Polyisocyanate compounds used for curing azido polymers.

The reaction between isocyanates and alcohols is catalyzed by organometallics (e.g., organotin compounds) and tertiary amines (triethanolamine). Dibutyltin dilaurate (DBTDL) is known to be a suitable catalyst for the formation of urethane between long chain diols and isocyanates.

2.4.1.2 Structure of Cured Polyurethane Elastomer

The structural repeating unit of the elastomer consists of the energetic polyether soft block linked to a hard urethane block. The alternating hard and soft blocks are the key feature that distinguish polyurethanes from other elastomer materials and is the primary reason that these materials have good mechanical properties over a wide temperature range [25].

Polyurethane elastomers utilize both physical and chemical cross-linking mechanisms in their constitution. Chemical cross-linking is the result of using a chain extension agent, whilst the physical cross-links are derived from the thermodynamic incompatibility of the non-polar energetic polyether soft blocks and the polar urethane hard blocks. The incompatible soft and hard blocks segregate into micro domains of soft and hard phases (Figure 2.3). The rigid urethane segments are held together in the discrete domains through Van der Waal's forces and hydrogen bonded interactions [26]. The rigid domains formed, act both as tied down points and as filler particles, reinforcing the soft segment matrix.

The final physical properties of the polyurethane elastomers depend upon the choice of appropriate polyisocyanate curing agents, the segment sizes of the soft

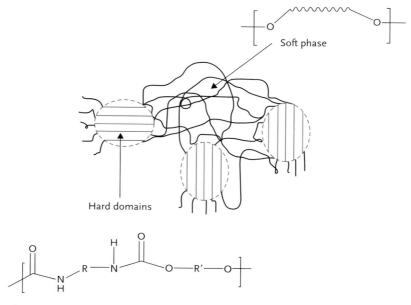

Figure 2.3 Schematic representation of the two-phase structure of the polyurethane elastomer.

and hard blocks, the nature of the chain extension agent, and the reactive ratio of isocyanate to the hydroxyl group.

2.4.1.3 Kinetics of Curing of Azido Polymers

Quantitative FTIR (Fourier Transform Infrared) spectroscopy is a useful tool to study the cure reaction between isocyanates and alcohols. The reaction can be monitored by the gradual decrease in the intensity of the absorption band at ~2270 cm^{-1} for NCO stretching in the IR spectrum and, the simultaneous appearance of the -COO- stretching band at 1726 cm^{-1} for polyurethane. The disappearance of the NCO stretching band and the growth of the -COO- stretching band of urethane demonstrates the conversion of isocyanate into the urethane group. The C–H stretching band at 2930 cm^{-1}, which remains unaltered as a result of the reaction, is usually kept as the internal standard. The degree of conversion at a given time $C(t)$ is calculated as:

$$C(t) = \left[1 - \frac{(A_{2270}/A_{\text{internal standard}})_t}{(A_{2270}/A_{\text{internal standard}})_0}\right] \quad (2.1)$$

where A is the FTIR absorbance of the designated peaks. Along with FTIR spectroscopy, viscometry and NMR (nuclear magnetic resonance) spectroscopy are also used for investigating cure kinetics of polymer binders.

Kezkin and Ozkar [27] studied the kinetics of curing of the hydroxyl terminated azido polymer, GAP, with the polyisocyanate, Desmodur N-100, using quantitative IR spectroscopy. The curing reaction obeys second-order kinetics until 50% conversion. At higher conversions, the reaction is controlled by the diffusion of reactants due to the high viscosity of the reaction mixture. It is premised that the shift from kinetic control to a diffusion controlled regime causes the deviation from second-order kinetics.

Ninan and coworkers [28] used FTIR spectroscopy to investigate the cure kinetics of GAP with TDI and IPDI. Similar to Desmodur N-100, the cure reactions follow a second-order rate law until 50% conversion. The cure rate is faster for TDI compared with IPDI.

It is interesting to compare the cure kinetics of GAP with that of inert polymer HTPB cured with TDI and IPDI, respectively. The curing reactions of HTPB with TDI and IPDI obey second-order kinetics with respect to the individual reactants [29–32]. The activation energies of uncatalyzed curing of GAP and HTPB polymers with TDI and IPDI are compared in Table 2.3.

Table 2.3 Comparison of activation energies of curing for GAP and HTPB.

Binder	Curing agent	Activation energy (kJ/mol)	Measurement technique	Reference
GAP	TDI	34.4	FTIR	[28]
HTPB	TDI	30.5	Viscometry	[29]
GAP	IPDI	73.9	FTIR	[28]
HTPB	IPDI	43.8	FTIR	[30]
HTPB	IPDI	41.0	Chemical analysis	[31]

It can be observed from the table that the activation energy of GAP curing is higher than that for HTPB with the same curing agent due to the low reactivity of secondary hydroxyl groups in the GAP polymer. Therefore, as compared with HTPB, curing of GAP is a slow process. Usually, depending on the NCO : OH ratio, it takes about 2–3 weeks to achieve the complete cure of GAP.

Catalysts are required to achieve curing of GAP in a reasonable time. Organotin catalysts such as DBTDL are very effective for catalyzing the reaction between alcohol and isocyanates, because they activate alcohols preferentially over water [33]. The proposed mechanism [34] is the reaction of the tin with the polyol, forming a tin alkoxide, which then reacts with isocyanate to form an N-coordinated complex. The alkoxide anion transfers to the coordinated isocyanate to produce N-stannylurethane, which undergoes alcoholysis to give the urethane and the original tin alkoxide (Scheme 2.7).

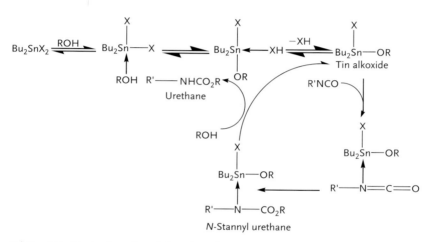

Scheme 2.7 Mechanism of catalytic action of DBTDL [34].

DBTDL catalyst, when dosed at 50 ppm, was found to enhance the rate of the curing reaction between GAP and Desmodur N-100 significantly [27]. Curing time reduces to a matter of 5–6 days from weeks.

Excessive gassing is a persistent problem with the curing of GAP formulations using DBTDL catalyst. Gassing has been attributed to a side reaction between the isocyanate curing agent and water [35]. The carbon dioxide produced as a result of the reaction gets trapped in the cross-linked structure causing voids in the cured binder, which can result in uncontrolled and unpredictable burn rates. A composite catalyst comprising of a mixture of triphenyl bismuth (TPB) and small amounts of DBTDL was found to effectively suppress the gas generation and prevent voiding [35]. Also, pot life and cure rate of the GAP/isocyanate mixtures could be tailored by varying the relative amount of DBTDL with respect to TPB in the composite catalyst system.

2.4.1.4 Gel-Time Characteristics

The uncured propellant slurry should have adequate pot life and reasonably low viscosity to enable casting of the mixture into a suitable configuration. Gel-time is a critical parameter determining the pot life of the uncured propellant formulation. Viscosity of the curing polymeric mixture increases with time due to formation of the urethane network. The point at which the viscosity becomes infinite is considered to be the gel-point [36]. The gel-point depends on the nature of the isocyanate, the OH : NCO ratio, and amount of curing catalyst. Gel-points are determined by viscosity build-up measurements. For GAP, at a given OH : NCO reactivity ratio, the viscosity build-up is faster for GAP/TDI system, followed by IPDI and methylene bis(cyclohexyl isocyanate) (MCHI) [28]. The viscosity build-up is slower for IPDI and MCHI due to the lower reactivity of the alicyclic isocyanates compared with the aromatic TDI. A similar trend was observed for reactivities of TDI and IPDI towards the inert HTPB [37]. Figure 2.4 compares the viscosity build-up profiles of GAP and HTPB during curing with TDI. The rate of increase of viscosity of the mixture is higher for HTPB compared with GAP, confirming the higher reactivity of HTPB over GAP.

For a given curing agent, Desmodur N-100, the gel-time of GAP is remarkably shortened by the addition of DBTDL catalyst to the reaction mixture [38]. As shown in Figure 2.5, gel-time of the mixture reduces by 800 min with a dosage of 40 ppm of catalyst.

Viscosity build-up profiles for the uncatalyzed and DBTDL catalyzed curing of HTPB with TDI are compared in Figure 2.6. The gel-time of HTPB polymer reduces by 140 min with the addition of the catalyst.

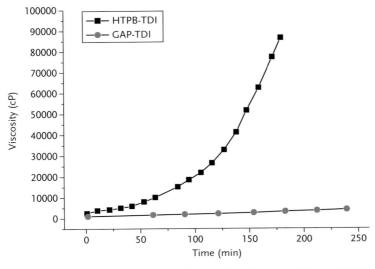

Figure 2.4 Comparison of viscosity build-up profiles for curing GAP and HTPB with TDI. Values of the plot are taken with permission of Springer from Refs. [28] and [37].

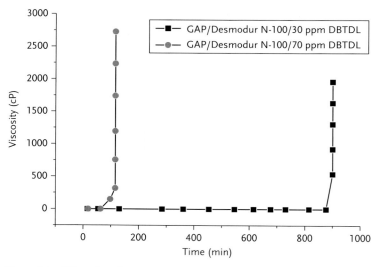

Figure 2.5 Effect of catalyst concentration on viscosity build-up for the GAP/Desmodur N-100 system. Values of the plot are taken with permission of Wiley from Ref. [38].

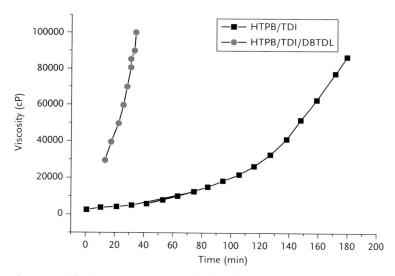

Figure 2.6 Effect of catalyst on viscosity build up for HTPB/TDI system. Values are taken with permission of Springer from Ref. [37]. Concentration of catalyst is 0.1% by mass of the polymer.

Comparing Figures 2.5 and 2.6, it can be observed that the gel-point during curing of the GAP polymer is more sensitive towards the addition of a catalyst as compared with HTPB. In other words, HTPB provides a wider processability window with respect to catalyst addition.

2.4.1.5 Post-Cure Properties

The final physio-mechanical properties of a polymer binder depend on the average molecular weight between the cross-links (M_c) in the cross-linked structure. M_c is a main characteristic parameter of the network, which is strongly affected by the nature of the cross-linking agent and the NCO:OH group ratio. In general, the higher the molecular weight across the cross-links (i.e., the longer the chain segments), the more elastic is the elastomer network. Eroglu and coworkers [39, 40] characterized the network structures of GAP and HTPB prepared by different curing agents, Desmodur N-100, IPDI, and Hexamethylene Diisocyanate (HMDI), through swelling measurements (Figure 2.7). The M_c values were obtained from the swelling measurements by applying the Flory–Rehner equation,

$$-[\ln(1 - v_2) + v_2 + \chi_1 v_2^2] = V_1 n [v_2^{1/3} - v_2/2] \tag{2.2}$$

where

v_2 is the volume fraction of the polymer in the swollen mass
V_1 is the molar volume of the solvent
n is the number of network chain segments bounded on both ends by cross-links
χ_1 is the Flory solvent–polymer interaction term.

HTPB formulations exhibit higher M_c values compared with GAP for a given curing agent and NCO:OH ratio. Evidently, the length of polymer segments between cross-link junctions is higher for HTPB as compared with GAP. Because

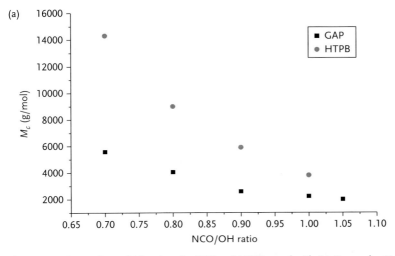

Figure 2.7 Comparison of M_c values for GAP and HTPB cured with (a) Desmodur N-100, (b) IPDI, and (c) HMDI. Values are taken with permission of Elsevier and Wiley from Refs. [39] and [40].

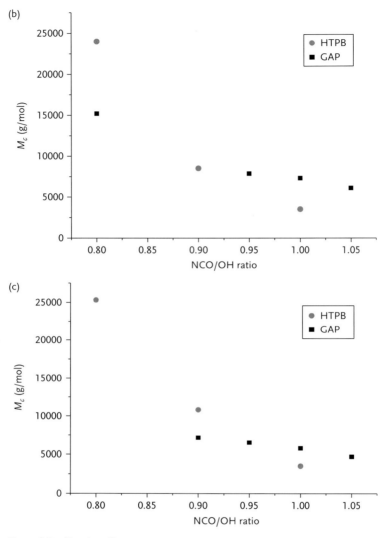

Figure 2.7 (Continued)

of this feature, the cured networks of HTPB are more elastic and exhibit higher values of elongation at break [40]. Among the curing agents, GAP and HTPB polymers cured with Desmodur N-100 have smaller values of M_c for the same NCO:OH ratio.

Another important post-cure property determining the performance of the polymer binder is the tensile strength. Tensile strength of the binder depends on the NCO:OH ratio and the nature of the curing agent, as illustrated in Figure 2.8.

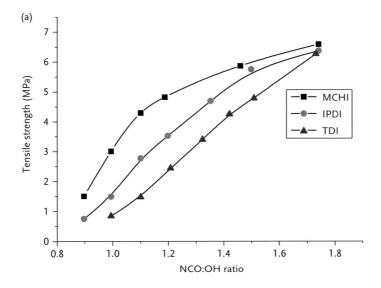

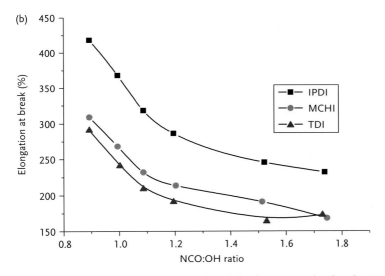

Figure 2.8 Variation of (a) tensile strength and (b) elongation at break with NCO:OH reactive ratio for GAP cross-linked with different curing agents. Values are taken with permission of Wiley from Ref. [41].

For a given curing agent, tensile strength of the cured GAP network increases with the NCO:OH ratio, with each of them eventually reaching a plateau level [42, 43]. With increasing NCO:OH ratio, the cross-linking density reaches a plateau level due to the formation of urethane–allophanate networks at higher NCO:OH ratios as explained in Scheme 2.8 [42].

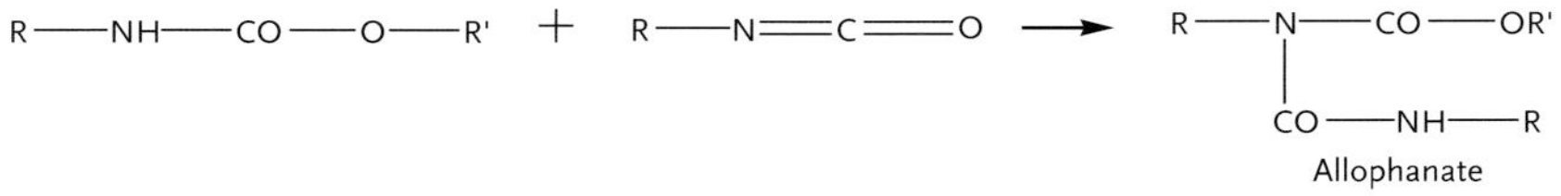

R—NH—CO—O—R' + R—N=C=O → R—N(—CO—OR')(—CO—NH—R)

Allophanate

Scheme 2.8 Mechanism of allophanate network formation.

At a given NCO:OH ratio, tensile strength increases in the following order of curing agents: MCHI > IPDI > TDI. Aliphatic diisocyanate curing agents (MCHI and IPDI) are capable of increasing the tensile strength of cured GAP more readily than the aromatic curing agent (TDI). The superior properties with aliphatic diisocyanates has been explained by systematic build-up of three-dimensional networks arising from the slow reactivity of the secondary OH groups of GAP with the aliphatic diisocyanate [41]. Elongations at break decrease with increasing NCO:OH ratio due to restricted inter-chain slippage imposed by the cross-links between the polymer chains.

Tensile strengths and elongation at break for HTPB binder compositions with various isocyanates are presented in Figure 2.9 for comparison with that of a cured GAP network.

In HTPB, the inter-chain forces are weak as the backbone contains hydrocarbons only. As a result, the tensile strength of cured HTPB is low [44]. With increasing NCO:OH ratio, a higher number of polyurethane bonds are formed between the chains, thereby increasing the tensile strength. In contrast with the case of GAP, the curing agent TDI is more capable of increasing the tensile strength of HTPB as compared with IPDI. Being an aromatic diisocyanate, TDI reacts more efficiently than IPDI with the primary hydroxyl groups of HTPB, causing a higher magnitude of the cross-linking density [42]. A higher cross-linking density of the HTPB–TDI composition restricts the inter-chain slippage thus leading to its higher tensile strength and lower elongation at break (Figure 2.9a and b) as compared with the HTPB–IPDI composition.

Glass transition temperature (T_g) is yet another important physical parameter that changes with curing. Curing restricts the segmental motions of polymer chains and shifts the T_g to higher temperatures. Such a shift will eventually lead to the hardening of the elastomer, thereby losing its elastic properties. The shift in the T_g of the GAP polymer has been studied with respect to the NCO:OH ratio during curing with Desmodur N-100 [38] and the results are presented in Figure 2.10.

It is interesting to note that a linear relationship exists between the NCO:OH ratio and T_g for the GAP polymer. Curing increases the T_g value of GAP by 6–9 °C depending on the NCO:OH ratio. Such a shift is detrimental to the functioning of GAP as a binder. Addition of plasticizers (i.e., a 1:1 mixture of bis(2,2-dinitropropyl) acetal (BDNPA) and bis(2,2-dinitropropyl) formal (BDNPF)) to the GAP polymer has been found to reduce the T_g of the binder to −46.7 °C, which is low enough to produce a rubbery propellant [38].

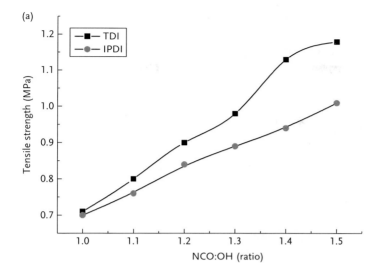

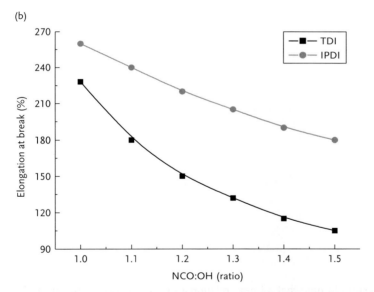

Figure 2.9 Variation of (a) tensile strength and (b) elongation at break for HTPB binder cured with TDI and IPDI, respectively. Values for the plot are taken with permission of Wiley from Ref. [42].

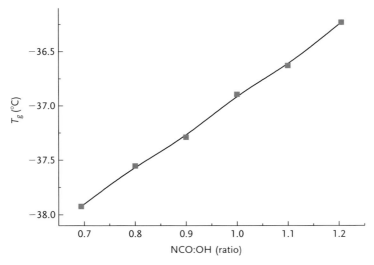

Figure 2.10 Variation of T_g with respect to the NCO/OH ratio during curing of GAP with Desmodur N-100. Values for the plot are taken with permission of Wiley from Ref. [38].

2.5
Curing of Azido Polymers by Dipolarophiles

The hydroxyl-functional azido polymers are often cured with polyisocyanates via a urethane-forming mechanism. However, this method of cross-link formation suffers from various shortcomings, such as: (i) risk of handling the isocyanates, (ii) interference by the moisture (gassing problem), and (iii) limitation due to stoichiometric factors. Hence, multifunctional dipolarophiles have been suggested as alternative curing agents for azido polymers [45]. Multifunctional dipolarophiles, comprising of esters or amides containing acrylic or acetylenic groups, react with azido groups of the polymers to form cross-linked polymer material containing triazole groups (Scheme 2.9).

The presence of triazole groups in the cross-linked polymer is expected to enhance the thermal stability and burn rate [46].

The extent of cross-linking can be tailored by varying the stoichiometric ratio of azido polymer to the dipolarophile. The cross-linked polymers belonging to various regimes exhibit different burn-rate behaviors. Table 2.4 compares the burn-rate behavior of dipolarophile cross-linked GAP polymer with that of the isocyanate cured one.

Data in the table show that cross-linking of GAP polyol with multifunctional dipolarophiles at dipolarophile azide stoichiometric ratios of about 0.2 : 1 or greater exhibited sustained burning. However, the isocyanate cured GAP polymers are self-extinguishing. Thus the triazole cross-linked polymers offer definite advantages over the isocyanate cured ones in their burn-rate behavior.

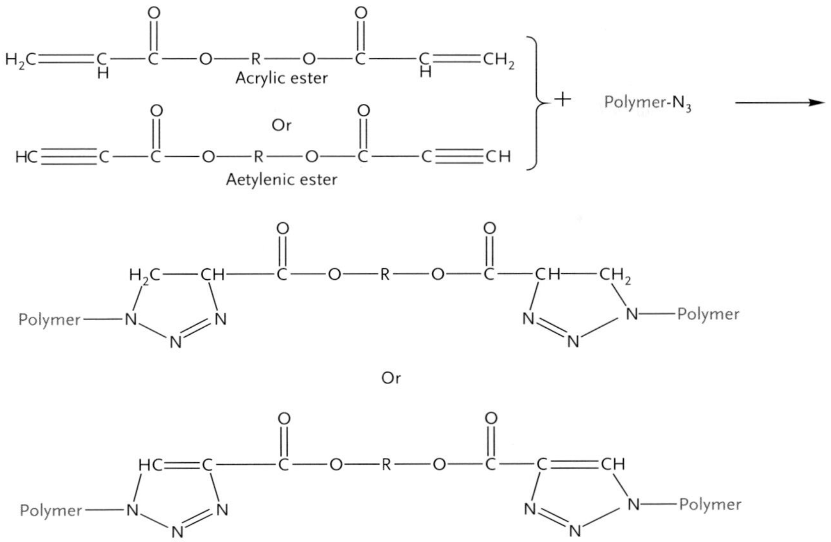

Scheme 2.9 Cross-linking of azido polymers using dipolarophiles.

Table 2.4 Comparison of burn-rate behavior of GAP polyol cross-linked with dipolarophile and isocyanate [46].

Cross-linking agent	Stoichiometric ratio (cross-linking agent : polymer)	Burn behavior
Dipolarophile	1:1	Sustained burning
Dipolarophile	0.60:1	Sustained burning
Dipolarophile	0.5:1	Sustained burning
Dipolarophile	0.4:1	Sustained burning
Dipolarophile	0.3:1	Sustained burning
Isocyanate (IPDI)	1:1	Self extinguishing
Isocyanate (N-3200)	1:1	Self extinguishing

However, the polymer obtained by Scheme 2.9 is inherently deficient as an energetic/elastomeric binder due to: (i) consumption of the energetic groups during the reaction (the triazole groups provide much less energy than the azide groups), (ii) insolubility of the polymer in energetic plasticizers, and (iii) rigidity of the polymer material [47].

A new synthetic route [48] (Scheme 2.10) has been developed in which the GAP is cured by reaction with diacetylenes to form triazole [49] cross-linked energetic binders having residual azido groups. Reaction times from 24 h to 1 week have

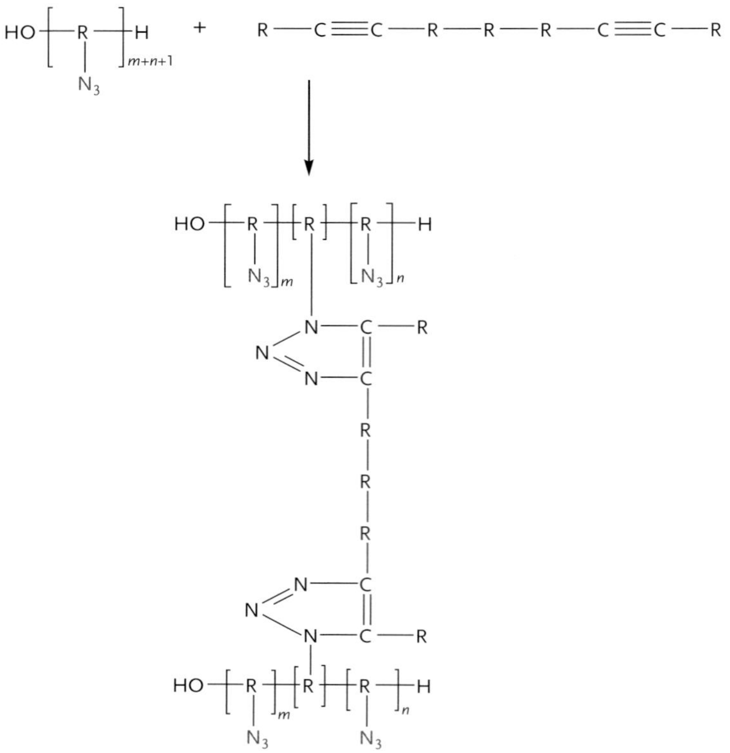

Scheme 2.10 Reaction scheme for the synthesis of triazole cured polymer with residual azide groups [48].

been reported for the completeness of the reaction due to the low reactivity of acetylenic compounds towards 1,3-dipolar cycloaddition reactions [50, 51].

Unlike the reaction of isocyanates with the hydroxyl groups to form urethane linkages, the triazole reaction needs no catalysts. The reaction rate is determined by the extent to which the acetylene bond is activated by any electron withdrawing groups in the vicinity of the molecule and by the ambient temperature. The activating groups are nitrites, carbonyls, ethylenes, and aryl moieties. With activated acetylenes, the curing will take place in the temperature range of 35–50 °C, which is practical for energetic material cure reactions [52]. During processing operations, the mixture is cast into a mold under vacuum. The mold is put into an oven for a week and cured into an elastomeric compound. Any leftover composition is stored in a container overnight at room temperature and added to another mix of a freshly prepared batch of the same composition next day. The leftover mixes stored at room temperature maintain a good correlation of burn rate with the freshly made batches. Thus, by using this methodology, the triazole cure materials can be recycled without degradation of performance properties [52].

2.6
Thermal Decomposition Characteristics of Azido Polymers

2.6.1
Mechanism of Thermal Decomposition

Over the past decade, several decomposition studies on azido polymers have been conducted using a variety of techniques. The techniques include SMATCH (simultaneous mass and temperature change)–FTIR [53], rapid thermolysis coupled with IR [54], T-jump–FTIR [55], thermal analysis [56, 57], mass spectrometry [58–60], and burning rate evaluation [61]. Results obtained from all these experiments demonstrate that the decomposition of azido polymer starts with the scission of the N–N_2 bond leading to the elimination of molecular nitrogen and formation of the corresponding polymeric imine or acrylonitrile (Scheme 2.11) [62]. The heat released during the thermal decomposition of GAP is due to the scission of the N_3 bond structure to form N_2.

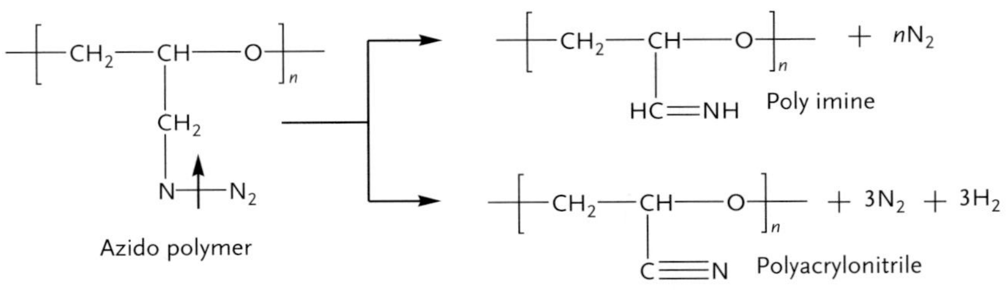

Scheme 2.11 Mechanism of thermal decomposition of the azido polymer GAP [62].

Once the azide structure is destroyed, electron shifting of the unpaired electron from the azide group occurs, resulting in the reduced stability of the remaining organic structure [63]. Upon continued heating, the molecular structure of the polymer undergoes further rupturing of aliphatic organic bonds with the release of various low molecular weight molecules. The final gaseous products of decomposition of the azido polymer consist of N_2, CO, HCN, NH_3, CH_2O, CH_4, C_2H_2, and oligomers of the azido polymer [58–60, 64, 65]. Relatively large molecules such as benzene, pyrrole, furan, polyatomic molecules containing nitrogen atoms, and free radicals have also been identified in the decomposition products [60]. In most of the decomposition and combustion experiments of azido polymers, solid residues have been observed [58, 61], which are high molecular weight polymers formed by the intramolecular cyclization or intermolecular cross-linking of the imino groups of the polyimine formed in the first step (Scheme 2.12).

The formation of the high molecular weight polymers have also been confirmed by FTIR spectroscopy, exhibiting peaks corresponding to the stretching vibrations of C=N–C–H or C=N–C–C in the molecule [66].

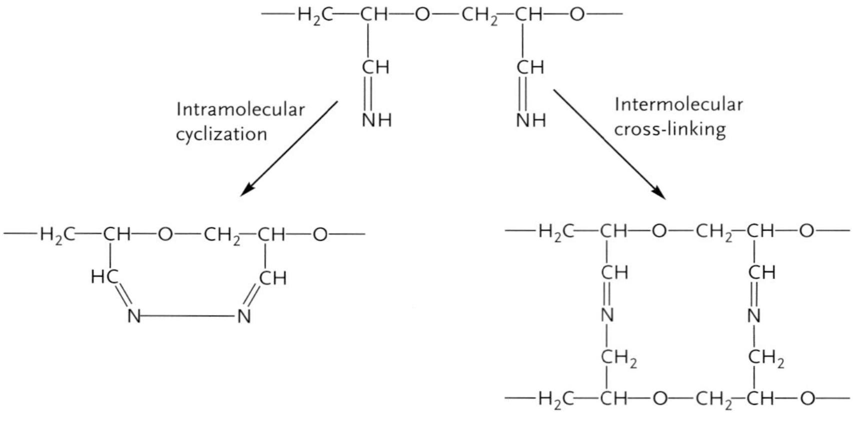

Scheme 2.12 Intermolecular cyclization and intermolecular cross-linking reactions of polyimine [59].

The decomposition mechanism of azido polymers is not significantly altered by heating rates and pressure differences [54]. The highly exothermic scission of the N_3 bond is the rate-determining step in azido polymers irrespective of the heating rates employed [62, 65]. However, the product distribution of GAP decomposition does depend on the experimental conditions employed [53, 54, 61, 63]. Ammonia is formed at lower heating rates and pressures. However, HCN is predominantly found at higher heating rates, relative to ammonia. Since HCN is toxic, attempts have been made to suppress the HCN content by using copolymers of GAP and isocyanate curing agents having higher hydrocarbon contents [57, 67].

Among the azido polymers (GAP, Poly(BAMO) and Poly(AMMO)), the composition of the decomposition products will vary due to the differences in their structures and energetics. A notable difference is observed for Poly(AMMO), which has the lowest enthalpy of decomposition among the three azido polymers. The decomposition product profile of Poly(AMMO) is dominated by formaldehyde (HCHO). The low molecular weight hydrocarbons found with GAP and Poly(BAMO) are absent in Poly(AMMO) [54]. Instead, high molecular weight 2-methylallyl compounds have been identified as decomposition products along with HCHO [53, 68]. GAP and Poly(BAMO) release more heat than Poly(AMMO), hence reactive molecules such as HCHO are immediately broken down into smaller molecules. The backbone structure of Poly(BAMO) starts to decompose simultaneously with scission of the azide bonds [68, 69] due to evolved heat. On the other hand, the backbone of Poly(AMMO) fails to pyrolyze completely even under severe conditions due to its lower enthalpy of decomposition. Therefore, Poly(AMMO) exhibits higher thermal stability compared with GAP and Poly(BAMO) [53, 63].

At this stage, it is interesting to compare the thermal decomposition of azido polymers with HTPB. HTPB decomposes in two stages [70–73] driven by a free-radical mechanism, which has been supported by electron spin resonance (ESR) studies [74]. At temperatures below 200 °C, the mechanism is initiated by H-abstraction resulting in the formation of free radicals. The resonance stabilized free radicals thus formed undergo cyclization or cross-linking reactions (Stage 1). In Stage 2, at temperatures above 450 °C, the cyclized and cross-linked polybutadienes formed in the first stage depolymerize to a complex mixture of hydrocarbons, namely butadiene, vinylcyclohexene, 1,3-cyclohexadiene, and cyclopentene. The cross-linking and cyclization reactions are exothermic, whereas depolymerization reactions are endothermic in nature.

Unlike the situation with azido polymers, thermal decomposition of HTPB is highly sensitive to heating rates. At higher heating rates, HTPB decomposes directly by the depolymerization mechanism as there is very little time for the exothermic cross-linking and cyclization reactions to occur [71, 75]. Hence no initial exotherm is observed in the reaction. The absence of an exotherm and the appearance of clipped polymer fragments (butadiene, vinylcyclohexene, which is a dimer of butadiene) in the decomposition products suggest that the combustion of HTPB binders is controlled by desorption of monomer fragments [75, 76] and not by any chemical bond breaking process as in the case of azido polymers.

2.6.2
Kinetics of Thermal Decomposition of Azido Polymers

The most common experimental techniques employed to study the kinetics of thermally activated reactions are thermogravimetry (TG), differential scanning calorimetry (DSC), and differential thermal analysis (DTA) [77]. These techniques are coupled with chemically specific detection methods such as FTIR, mass spectrometry (MS), and gas chromatography (GC), to analyze the gaseous products formed from the decomposition reaction. Whereas TG, DSC, and DTA provide information about the global kinetics of thermally stimulated reactions in terms of kinetic parameters; activation energy (E) and frequency factor (A) [78, 79], the complementary techniques permit a deeper insight into reaction mechanisms [80, 81]. These methods were successful in the study of the binder pyrolysis under mild conditions. However, doubts exist over the extrapolation of thermal decomposition results for the binder obtained under mild conditions to the severe conditions of propellant combustion, where the magnitude of heat flux (10^6 K/s), temperature (2000–3500 K), and pressure (5–10 MPa) encountered by the binder are very different [74, 76]. Over the years, fast pyrolysis techniques such as SMATCH (simultaneous mass and temperature change)–FTIR and T-jump–FTIR spectroscopies have emerged, which can correlate the microscale fast thermolysis with the steady-state combustion of the bulk materials [53, 68].

TGA and DSC curves of representative azido polymers (GAP and Poly(BAMO)) presented in Figure 2.11 [82], show that the main weight loss step in TGA coincides with the exothermic decomposition peak in the DSC curve. The weight loss is due to the exothermic scission of azido groups to release nitrogen, the main decomposition mechanism of azido based energetic polymers.

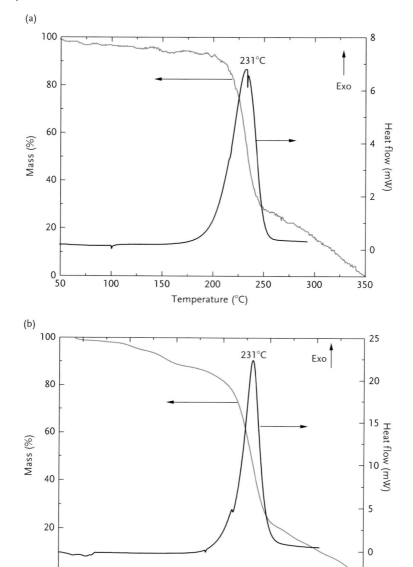

Figure 2.11 Overlay of TGA and DSC curves for (a) GAP and (b) Poly(BAMO). Plots are taken with permission of Elsevier from Ref. [82].

2.6 Thermal Decomposition Characteristics of Azido Polymers

The enthalpy of decomposition and magnitude of weight losses depend on the number of azido groups in the polymer. Poly(BAMO) decomposes with a higher enthalpy of decomposition of 1970 J/g as compared with GAP (1696 J/g) due to the higher number of azido groups in Poly(BAMO). Also, due to the same reason, Poly(BAMO) exhibits a higher weight loss of 54% as compared with GAP, which shows a 42% weight loss in the temperature range of 200–250 °C.

Activation energies of thermal decomposition of the azido polymers have been calculated from the Arrhenius plots (Figure 2.12) generated from the isothermal TGA measurements assuming first-order reaction kinetics [53].

From the Arrhenius plots, the order of rates of isothermal decomposition is observed to be: GAP>BAMO>AMMO (Table 2.5).

All the azido polymers have activation energies in the range of 164–172 kJ/mol. This result is conclusive evidence to the fact that azido polymers, irrespective of their difference in structure, decompose primarily by the scission of the azide bond.

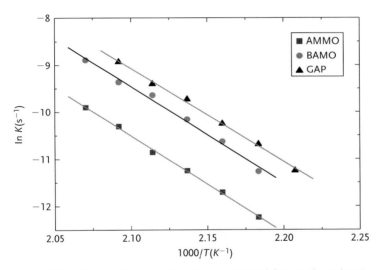

Figure 2.12 Arrhenius plots for azido polymers obtained from isothermal TGA measurements. Values for the plot are taken with permission of Elsevier from Ref. [53].

Table 2.5 Arrhenius parameters for different azido polymers [53].

Azido polymer	Activation energy (kJ/mol)	Ln A (1/s)
AMMO	171.1	32.73
BAMO	167.12	33.21
GAP	164.81	32.56

Table 2.6 Arrhenius parameters for thermal decomposition of different types of GAP.

Technique	E (kJ/mol)	A (1/s)	Remark	Reference
DTA, TGA	174.2	Not mentioned	Cured GAP	Kubota et al. [61]
TGA	164.92	4×10^{14}	Uncured GAP Diol	Chen et al. [53]
DSC	160	1.25×10^9	Uncured GAP Triol	Jones et al. [83]
DSC	138	3.73×10^{11}	Branched GAP	Feng et al. [84]
SMATCH (Simulated Combustion)	180.9	1.0×10^{19}	Uncured GAP Diol	Chen et al. [53]

Among the azido polymers, thermal decomposition of GAP has been extensively studied. The Arrhenius activation energies of thermal decomposition of uncrosslinked and cured GAP determined by various researchers using different techniques are summarized in Table 2.6.

The values are mostly in the range of 135–181 kJ/mol, which is similar to the strength of the RN–N$_2$ bond in alkyl azides. However, some differences are observed when the macromolecular architecture of the polymer changes from linear to branched. The activation energy of branched GAP is lower than that for GAP Triol, indicating that branched GAP is more reactive [83, 84]. Another important point is that the activation energies for thermal decomposition of GAP measured at slower and faster heating rates are similar. This means that the decomposition mechanism of azido polymer does not vary with heat rate. However, the heat of reaction evolved in a fast-rate decomposition (−5.6 kJ/g) is larger than the heat released upon the slow rate decomposition of GAP measured by DSC (−1.7 kJ/g) [55].

TGA and DSC curves of HTPB resin heated at 5 °C/min up to 700 °C are presented in Figure 2.13.

HTPB shows two major weight loss stages with indistinct separation, as also reported earlier [73]. At this heating rate, the first stage exhibits a much smaller weight loss, of 19.2%, than the second stage (76.7%). The DSC profile shows that the first stage is exothermic, with a net heat release of 700 J/g. The exothermicity is because of the prevalence of the energy releasing cross-linking and cyclization processes of HTPB over that of the bond breaking depolymerization [75]. The second stage weight loss is the result of decomposition of the cross-linked residue formed by cyclization and cross-linking reactions in the first step [85]. This step appears as an endothermic transition in the DSC curve ascribed to the depolymerization of residue formed in the first step.

As mentioned in the previous section, the thermal decomposition of HTPB is susceptible to heat rates, with the mechanism shifting from cyclization–cross-linking at low heat rates to depolymerization at higher rates [71, 75, 85]. Hence, unlike the case with azido polymers, the activation energies of thermal decomposition of HTPB polymer vary with heat rates (Table 2.7).

It is interesting to note that in contrast to azido polymers, for HTPB there is relatively poor agreement between the Arrhenius parameter data at low heating rates. This suggests that the depolymerization–cross-linking reactions are

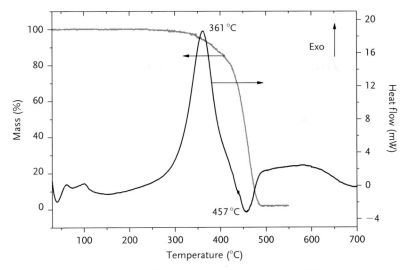

Figure 2.13 TGA and DSC profiles for HTPB polymer heated at 5 °C/min heating rate. Unpublished results (Pisharath and Ang).

Table 2.7 Arrhenius parameters for HTPB at slow and fast heating rates.

Technique	E (kJ/mol)	Ln A (1/s)	T (°C)	Reference
TGA	78.71	2.5	328–420	Chen et al. [75]
TGA	115.56	16.2	350–400	Du et al. [85]
TGA	251.627	17.1	367–407	Du et al. [85]
SMATCH (simulated combustion)	34.8	7.13	–	Chen et al. [75]

occurring in parallel ways. Another important difference to note is that the activation energy of thermal decomposition of HTPB at a higher heat rate is significantly smaller than at lower heating rates. The interpretation of a low value for the activation energy is that at high heat rates the kinetics are controlled by the rate of desorption of the species from the surface as opposed to the rate of bond-breaking at slow heat rates in the bulk phase [75, 76]. For the azido polymers, however, the kinetics of thermal decomposition are controlled by the scission of the azido group at both slower and faster heat rates.

2.7
Combustion of Azido Polymers

Combustion mechanisms of azido polymers have been described by a number of thermo-chemical and burn-rate studies [53, 58, 86–90]. A schematic diagram of the thermal profile during combustion of azido polymers is presented in Figure 2.14.

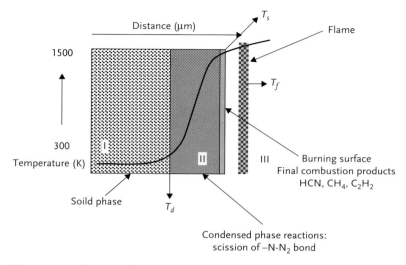

Figure 2.14 Schematic of thermal profile during the combustion of energetic azido polymer.

The thermal profile during combustion of the azido polymer can be divided into three zones. Zone I is a non-reactive heat conduction zone, zone II is a condensed phase reaction zone, and zone III is a gas phase reaction zone in which the final combustion products are formed. The decomposition reaction of the azide group occurs at the temperature denoted by T_d in zone II, and the gasification reaction completes at the temperature denoted by T_s (surface temperature) at the end of zone II. Very high values of T_s have been reported for azido polymers, which are measured with Pt–Pt/Rh thermocouples. The measured values are in the range of 700–800 K for pressures from 0.4 to 10 MPa [86, 89, 90]. High surface temperature values suggest that the dominant combustion reaction of azido polymers (i.e., scission of the –N–N$_2$ bond of the azido group) proceeds in the condensed phase. Many workers [90, 91] have reported the formation of a carbonaceous residue on the combustion surface due to the fuel-rich nature of the azido polymers.

The heat balance equation on the combustion surface is represented as [92].

$$\rho_p c_p r(T_s - T_0) = \lambda_g \left(\frac{dT}{dx}\right)_g + \rho_p r Q_s \qquad (2.3)$$

where

ρ is the density of the polymer
c is the specific heat
r is the burn rate
T_s is the temperature on the combustion surface
T_0 is the initial temperature
x is the distance
λ_g is the thermal conductivity of the gas phase
Q_s is the heat generation on the combustion surface.

Table 2.8 Comparison of burn rates for various polymers [93].

Polymer	Burn rate (mm/s) (obtained by combustion measurement)
GAP	1.7
Nitrocellulose	0.4
HTPB	0.19

Equation (2.3) could be reduced to

$$r = \frac{\lambda_g \left(\frac{dT}{dx}\right)_g}{\rho_p \{c_p(T_s - T_0) - Q_s\}} \quad (2.4)$$

It is clear from Eq. (2.4) that the burn rate is dependent on two parameters: the heat feedback from the gas phase determined by λ_g, and by Q_s, the heat generation on the combustion surface. Azido polymers exhibit high burn rates compared with inert polymer (HTPB) and other energetic binders, such as nitrocellulose (NC), due to the large heat generation Q_s on the combustion surface by the rupture of the $-N-N_2$ bond of the azido group (Table 2.8). The heat flux transferred back from the gas phase to the condensed phase is very small compared with the heat generation from the combustion surface [86].

The burn rate of azido polymers shows a positive linear correlation with the azide content of the polymer because the main heat source of combustion is the decomposition of the azido groups [88, 65, 94]. For example, the burn rate of Poly AMMO (having the lowest heat of decomposition among the three azido polymers) is 50% that of the GAP binder [89]. The large effect of the azide content points to the significant impact of the condensed phase energy release.

The flame temperature of any polymer fuel is determined by the heat evolution in the gas phase (zone III in Figure 2.13) due to the reaction between the combustion products. In azido polymers, as the combustion products are fuel rich and chemically inert, the only possible source of heat generation is from the decomposition of depolymerized fragments of the azide polymer, which would have escaped to the gas phase from the condensed phase by means of evaporation [95]. Hence, azido polymers produce lower flame temperatures (~1300–1400 K) [96], significantly lower than that for the energetic polymer NC (~2344 K) [94].

2.8
Thermal Decomposition and Combustion of Energetic Formulations with Azido Polymers

The discussion in this section will address the thermal decomposition and combustion aspects of azido polymer based formulations containing different oxidizers.

2.8.1
Thermal Decomposition of Azido Polymer/Nitramine Mixtures

There is considerable interest in the use of high performance nitramine oxidizers in gun and rocket propellants due to their desirable properties, such as absence of noxious combustion products, high specific impulse, and thermal stability. However, nitramine oxidizers suffer from low burn rate characteristics, high pressure rate exponents, and very high flame temperatures (3000 K). Therefore, nitramine explosives are mixed with fast burning rate azido polymers to tailor the properties of the explosive. Along with the burn-rate characteristics, the kinetic parameters, that is, activation energy and the pre-exponential factor for thermal decomposition, are altered by mixing azido polymers with nitramines. Kinetic parameters are useful for safety evaluation and performance prediction of energetic formulations [97].

Extensive studies have been carried out on individual ingredients of nitramine/azidopolymer mixtures; however, despite the increased interest, only a limited numbers of results are available on the decomposition characteristics of mixtures.

In this section, the effect of mixing of azido polymers on the thermal decomposition behavior of nitramines is discussed.

Oyumi et al. [98] studied the thermal decomposition of HMX/BAMO propellants using isothermal TGA and DSC. It was shown that the decomposition of HMX was initiated and accelerated by the exothermic decomposition of BAMO binder in the propellant.

Activation energy for thermal decomposition of different nitramine/GAP mixtures have been evaluated experimentally by Ger et al. [99] and computationally by Bohn and coworkers [100]. The values are compared with those of pure molecules in Table 2.9.

The primary mechanism in thermal decomposition of nitramines is the homolytic fission of $N-NO_2$ bonds in the molecule [101]. The activation energy for thermal decomposition of an RDX/GAP mixture is higher than pure RDX and the values are comparable for an HMX/GAP mixture and pure HMX. On the other hand, the activation energy for a CL-20/GAP mixture is lower than pure CL-20, suggesting that the $N-NO_2$ bond cleavage in CL-20 is energetically favored in the presence of GAP as compared with that in HMX and RDX.

Table 2.9 Comparison of activation energies of nitramines and their mixtures with GAP.

Nitramine	Activation energy of pure nitramine (kJ/mol)	Activation energy of GAP/nitramine mixture (kJ/mol)	Reference
HMX	185[1]	185.1[a]	[99, 100]
RDX	171[1]	178.9[a]	[99, 100]
CL-20	153.3	148.6	[100]

[a]Average of values taken from Refs. [99] and [100].

2.8 Thermal Decomposition and Combustion of Energetic Formulations with Azido Polymers

TEX (4,10-dinitro-2,6,8,12-tetraoxa-4,10-diazaisowurtzitane) is another high performance nitramine oxidizer with high detonation velocity and low sensitivity towards friction and impact [102]. At our Institute, the thermal decomposition characteristics of a TEX/GAP mixture were investigated in detail using DSC and TGA and the kinetics evaluated by employing an isoconversional method [103].

In contrast to other studies on GAP/nitramine mixtures [99], the decomposition temperature of TEX does not change when mixed with GAP, as shown by the DSC profiles in Figure 2.15.

The DSC curves indicate that the decomposition of the mixture is not controlled by any molecular level interaction between GAP and TEX.

Thermal decomposition kinetics of the GAP/TEX mixture was investigated by an isoconversional method, which provides a model-free estimate of activation energy as a function of the extent of decomposition. The resulting dependency of activation energy on the extent of decomposition provides information on the mechanistic aspects of the decomposition process.

The dependencies of activation energy on the extent of decomposition shown in Figure 2.16 illustrate that the decomposition of the pure molecules are controlled by single-step mechanisms.

On the other hand, for the GAP/TEX mixture, the dependence curve exhibits a decreasing trend from 110 ± 2 to 106 ± 2 kJ/mol in the conversion range of $0.15 \leq \alpha \leq 0.25$ (see inset in Figure 2.16) suggesting that the exothermic decomposition of GAP is competing with the solid-state decomposition of TEX in

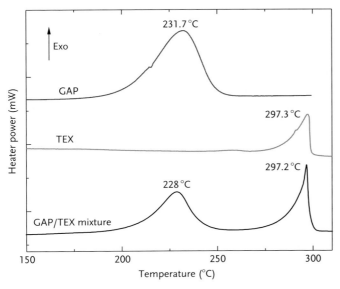

Figure 2.15 DSC curves for GAP, TEX and GAP/TEX mixture. Plot taken from Ref. [103] with permission from Elsevier.

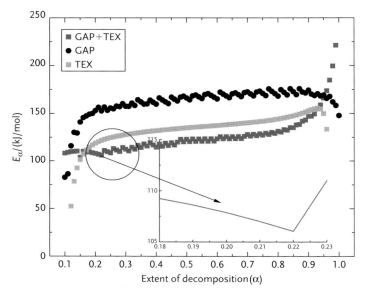

Figure 2.16 Dependencies of activation energy on the extent of decomposition for GAP, TEX, and GAP/TEX mixture. Plot taken with permission of Elsevier from Ref. [103].

this conversion range. The activation energy profile of the GAP/TEX mixture is lower than that of TEX in the conversion range of $0.25 \leq \alpha \leq 0.98$.

As shown in Figure 2.17, a combination of hot stage optical microscopy (OM) and Fourier transform infrared (FTIR) can be used to investigate the decomposition of the GAP/TEX mixture. In the experiment, the mixture was kept on the hot stage of the optical microscope and the mixture was heated at a controlled heating rate of 2 °C/min from room temperature to 300 °C. The morphological changes of the mixture during heating were observed. The residues of the mixture were collected at specific temperatures and FTIR of the residues were taken by dispersing them on KBr pellets.

The room temperature optical micrograph of the GAP/TEX mixture shows a fine coating of GAP over the TEX particle. The FTIR spectrum of the mixture shows characteristic peaks due to the azide group ($-N_3$) stretching in the GAP polymer at 2102 cm^{-1} and N–NO$_2$ stretching peak for TEX at 1595.02 cm^{-1}. Upon heating, the GAP component in the mixture decomposes at 238 °C, which is reflected in the FTIR spectrum as the absence of the peak due to the azide. The exothermic decomposition of GAP induces porosity on TEX, as illustrated by the optical micrograph of the GAP/TEX mixture. The porous structure of TEX promotes easier diffusion/adsorption of the hot decomposition products of GAP resulting in the decomposition of TEX. The exothermic decomposition of TEX in the GAP/TEX mixture is thus controlled by the mass transport along the solid surface, which is characterized by relatively lower

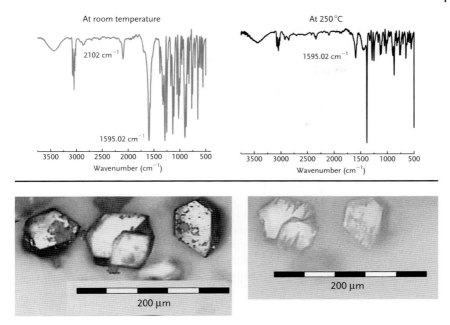

Figure 2.17 Comparison of optical micrographs and FTIR spectra of GAP/TEX mixture at room temperature and at 250 °C. Plot taken with permission of Elsevier from Ref. [103].

activation energy compared with the kinetic controlled unimolecular decompositions of GAP or TEX.

The kinetic scheme thus derived from the experimental observations (Figure 2.18) is necessary for a fundamental understanding of the thermal decomposition process of azido polymer/nitramine mixtures.

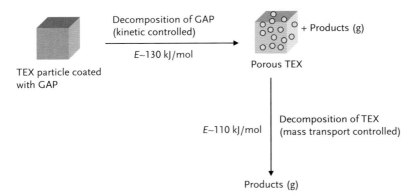

Figure 2.18 Kinetic schemes for the thermal decomposition of GAP/TEX mixture.

2.8.2
Combustion of Azido Polymer/Nitramine Propellant Mixtures

Mixtures of propellant ingredients exhibit different combustion characteristics to those of the constituent individual ingredients. The major factor contributing to this phenomenon is the difference in chemical kinetics, which affects the spatial distribution of the energy release [104]. In general, the burn rate of nitramines decreases with the addition of the azido polymer GAP [105–108], even though GAP burns faster than the nitramines. This is attributed to the dilution in the concentration of the reactive species released from the decomposition of the nitramines by the nitrogen released from the decomposition of the GAP. Consequently, the heat feed back from the gaseous-phase reactions to the burning surface decreases, resulting in reduced burn rate of GAP/nitramine mixtures [104–107]. This feature is exemplified by the burn-rate characteristics of the GAP/HMX mixture and the individual components, shown in Figure 2.19. It is to be noted that there is a significant lowering of the pressure exponent of HMX on mixing it with GAP, which improves the pressure sensitivity of the propellant.

In order to comprehend the combustion characteristics of nitramine/azide propellant mixtures, an understanding of the combustion wave structure is necessary (Figure 2.20) [109].

The combustion wave structure of the GAP/nitramine propellant mixture consists of a primary (first stage) reaction zone, a dark zone (also known as the preparation zone), and a second stage reaction zone [110]. The dark zone is a

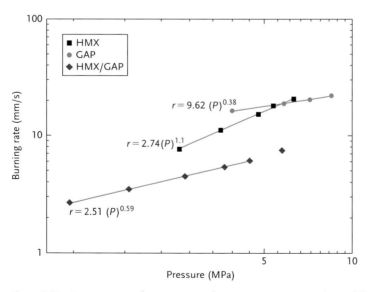

Figure 2.19 Burning rates of HMX, GAP and GAP/HMX mixtures. Values of the plot are taken with permission of Wiley from Ref. [106].

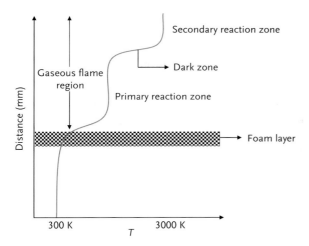

Figure 2.20 Schematic of combustion wave structure of GAP/nitramine propellant mixture.

non-luminous region between the primary and secondary flame zones with a temperature plateau [109].

The reaction zone chemistry in the combustion of the GAP/nitramine mixture is dominated by nitramines. Initial decomposition pathways of nitramines producing NO_2 and HCHO are:

$$HMX_{(1)}(RDX)_{(1)} \rightarrow 4(3)HCHO + 4(3)N_2O$$
$$HMX_{(1)}(RDX)_{(1)} \rightarrow 4(3)HCN + 2(1.5)(NO_2 + NO + H_2O)$$

The formaldehyde forming reaction is exothermic (favored at low heat-rate conditions) and the second one, initiated through the N–NO_2 cleavage, generating HCN, is endothermic in nature.

The primary reaction zone is associated with the reaction of nitrogen dioxide (NO_2) with formaldehyde (HCHO) taking place at the subsurface multiphase region, called the foam layer. This region contains a liquid layer of nitramine with gas bubbles and other condensed species. The burn rate of the mixture is controlled by the reactions that take place in the primary zone.

The other intermediates (HCN, NO, CO, N_2O) produced in the primary zone require higher activation energies (~189 kJ/mol) to react amongst each other. In order to overcome the high activation energy barrier, the reactants should have a finite residence time, within which the species may travel a certain distance before reacting at the secondary reaction zone. During the finite residence time, the variations in temperature and species concentration are moderate and no visible emission is observed in this zone, thus referred to as the dark zone [110].

The dominant reaction in the secondary reaction zone is that between HCN and NO [105] to release N_2 and CO:

$$2HCN + 2NO \rightarrow 2CO + 2N_2 + H_2$$

The influence of the azido polymer on the combustion of the GAP/nitramine mixture is to influence the primary reaction zone length and also the location at which the secondary reaction zone begins [105]. Energetic polymer addition also changes the relative surface molecular fractions of some of the species generated with nitramine combustion, including HCHO, N_2, and HCN, because these are the major products generated during the decomposition of azido polymers. In addition, carbon dioxide is observed at the propellant surface, which is not present at the surface of the neat nitramines [105].

The problem with GAP/nitramine propellants is their high pressure exponent values, which make them unsuitable to be used in rocket motors [111]. The pressure sensitivity could be lowered to practical levels by the addition of appropriate catalysts (Figure 2.21).

In the case of GAP/HMX propellant, a mixture of lead citrate and carbon black catalyst yields a burning-rate law with moderate pressure exponent and superior burning rate [110]. The burn rate is controlled by the heat released from the burning surface. The flame stand-off distance increases when the propellant is catalyzed due to the enhancement in the flow velocity of the gas.

Usually propellant formulations containing GAP and nitramines are fuel rich and hence burn poorly, forming a carbonaceous residue as they burn. Therefore, addition of a highly oxygenated plasticizer, butanetriol trinitrate (BTTN), to GAP/nitramine mixtures provides more oxygen and thus reduces char formation. Combustion of ternary mixtures of GAP/nitramine/BTTN has been studied in the literature [112–116]. As demonstrated in Figure 2.22, addition of BTTN to a

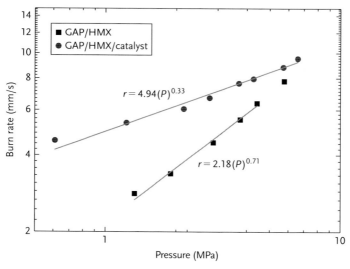

Figure 2.21 Effect of catalyst (lead citrate + carbon black) on the burn rate of GAP/HMX propellant. Values for the plot are taken with permission of Elsevier from Ref. [110].

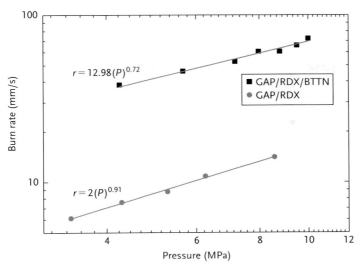

Figure 2.22 Effect of addition of BTTN plasticizer on the burn rate of GAP/RDX mixture. Values for the plots are taken with permission of Elsevier from Refs. [108] and [112].

GAP/RDX mixture significantly improves the burn rate and also decreases the pressure exponent values.

Addition of oxygen-rich BTTN helps to balance the fuel-rich nature of GAP/RDX mixtures stoichiometrically, leading to more heat release near the burning surface, which consequently increases the burn rate of the mixture [104, 112]. Also, combustion of this mixture has been found to be relatively free from carbonaceous residues and smoke [115]. As in the case of GAP/nitramine propellants, the GAP/nitramine/BTTN formulation also shows a dark zone in the flame structure where the concentrations of NO and HCN are at their highest.

SNPE has developed experimental propellant formulations based on CL-20 and GAP with impressive burn rates and an acceptable pressure exponent [117]. As shown in Figure 2.23, ballistically modified GAP/CL-20(40:60) propellant formulation have an impressive burn rate in the range of 20–32.4 mm/s in a pressure range of 7–20 MPa with a pressure exponent of 0.46. The burn rate at 15 MPa of the GAP/CL-20 propellant is 20% higher than that of the inert binder propellant.

It is clear that among the nitramines, CL-20 provides a strong effect on burn rates when used as an oxidizer in propellant.

2.8.3
Combustion of Azido Polymer/Ammonium Nitrate Composite Propellants

Ammonium nitrate (AN) is an environmentally friendly oxidizer that produces clean burning propellants. Of late, with the increasing interest for environmentally friendly chlorine-free propellants, there is growing interest in AN propellants,

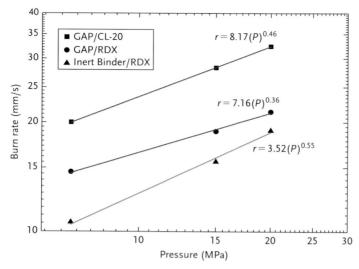

Figure 2.23 Comparison of burn-rate profiles of ballistically modified GAP propellants of CL-20 and RDX with an inert binder/RDX propellant. Values of the plot are taken with permission of Dr. Yves Longevialle from Ref. [117].

which produce harmless combustion products. However, their application has been limited to low-performance applications, such as gas generators, due to low burn rates and energy content [118].

The poor combustion performance of AN is expected to be improved by combining the oxidizer with the high burn rate energetic binder GAP, which will add in additional energy to the formulation. However, a GAP/AN formulation (70 : 30 weight ratio) was unable to sustain combustion below 4 MPa pressure [119].

As shown in Figure 2.24, energetic plasticizers, such as BTTN, TMETN, and nitrate esters, combined with chromium catalysts provide impressive burn rates to GAP/AN formulations [119, 120] and also reduce the pressure exponents. The burn rates could be improved further by employing molybdenum/vanadium oxide catalysts (MOVO). With MOVO, burn rates of the order of 7–8 mm/s and pressure exponents of 0.5–0.55 could be achieved for GAP/AN formulations combined with nitrate esters [121]. An optimal combination of burn rate and pressure exponent could also be achieved from the GAP/AN formulation by the addition of the crystalline energetic nitrate triaminoguanidine nitrate (TAGN) in the range of 20–30% mixed with a combination of organometallic bismuth, zirconium carbide, and carbon black [122].

2.8.4
Combustion of Azido Polymer Propellants with HNF

Hydrazinium nitroformate (HNF), $N_2H_5C(NO_2)_3$, holds promise as a clean burning, high performance energetic oxidizer having a positive oxygen balance (+13.1%) and substantially larger heat of formation (−72 kJ/mol) compared with

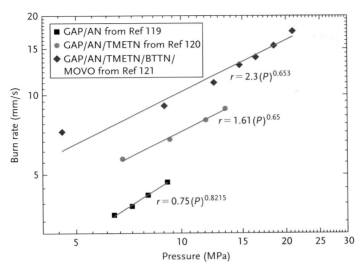

Figure 2.24 Comparison of burn rates for GAP/AN, GAP/TMETN/AN, and GAP/TMETN/BTTN/MOVO compositions. Values for the plot are taken with permissions of Elsevier and Wiley from Refs. [119–121].

AN (−365.04 kJ/mol) [122]. In comparison with ammonium perchlorate (AP), it is environmentally benign due to the absence of chlorine atoms in the molecular structure. In a research effort to develop high performance HNF propellants, the energetic azido polymer GAP has been suggested as the fuel for the formulation [123, 124]. Most of the HNF propellant research is being carried out in The Netherlands [125]. HNF decomposes quickly with a surplus of oxidizing species, H_2O and CO_2 [126]. These oxidizing species are expected to react with the fuel-rich GAP at the propellant surface, thereby increasing the burn rate of formulation. The burn rate of pure HNF is compared with that of GAP/HNF and HTPB/HNF propellants in Figure 2.25 [127–129].

Both the HNF propellants have lower burn rates as compared with the pure HNF. HNF burns very fast with a high pressure exponent (closer to one at high pressures) producing a reactive hot flame near to the propellant surface [128]. In GAP/HNF formulations, at lower pressures, the burn rate of the energetic binder is lower than that of HNF, and the HNF burns away leaving back the binder. In this situation, the burn rate of the propellant is solely controlled by the condensed-phase decomposition of HNF. However, at higher pressures, the burn rate of the binder catches up with the burn rate of HNF and forms a diffusion flame far from the propellant burning surface [130]. Thus GAP takes an active participation in the combustion of the GAP/HNF propellant. As the diffusion flame is formed at a great distance from the burning surface, the conductive heat transfer from the gas phase is not strong enough to become the controlling mechanism of GAP/HNF propellant burning [130]. Therefore, the burn rate of the GAP/HNF propellant is lower than that of pure HNF. In the HTPB/HNF propellant, the binder does not support the combustion of the propellant because of its inert nature, thus showing

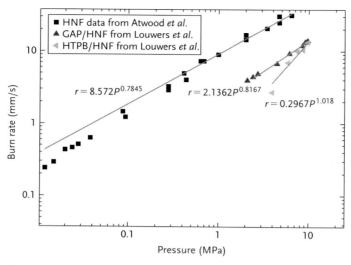

Figure 2.25 Burn-rate comparison of pure HNF with GAP/HNF and HTPB/HNF propellants. Plot values taken with permission of AIAA and Prof. Dirk Roekaerts (Delft University of Technology) from Refs. [127–129].

a lower regression rate compared with the GAP/HNF propellant [129]. Moreover, when HNF is used with unsaturated hydrocarbon binders, such as HTPB, the HNF attacks the double bond of the binder and breaks down the backbone of the binder, thereby softening and weakening the propellant [124, 131].

Both the propellants show values of pressure exponent that are far too high (0.81 for GAP/HNF and 1.08 for HTPB/HNF), which are unacceptable for the use of these formulations as propellants. Suitable burn-rate modifiers have been identified by the European Space Agency (ESA) that could reduce the pressure exponent to roughly below 0.6 [124, 132]. Another hurdle faced in HNF propellant development is the incompatibility of the isocyanate curing agents with GAP [133]. A possible reason for the incompatibility is the transfer of hydrogen from HNF to nitrogen in $-N=C=O$, which has been confirmed on the basis of an 88% decrease in the isocyanate absorption peak in the IR spectra of an HNF/isocyanate mixture over a period of 48 h [134].

New curing systems based on acrolyl chloride (AC) have been tested [124], in which the double bond of the chloride reacts with the azido group of GAP to produce a triazole cross-linked structure, as shown in Scheme 2.9. The new curing system offers the following advantages: (i) room temperature curing reactions, (ii) eliminates the possibility of the toxicity and moisture sensitivity of isocyanate based curing agents, (iii) allows the tailoring of mechanical and ballistic properties by varying the stochiometry of the reaction between the polymer and curing agent, and (iv) enhances thermal stability and burn rate due to the formation of triazole groups. Vacuum thermal stability (VTS) tests have indicated good compatibility between HNF and the new curing system [124]. Also, the formulations exhibit good ageing characteristics. Thus, the new curing system based on AC appears to be promising for GAP/HNF propellants. However, mechanical properties of the new formulations are yet to be assessed thoroughly in order for GAP/AC/HNF formulations to become a commercial reality.

2.9
Performance of Azido Polymer-Based Propellants

Performance of any rocket is determined by the rocket equation first derived by Konstantin E. Tsiolkovsky in 1903. According to the equation, the velocity increment of the rocket stage (Δv) is represented as [124, 135]:

$$\Delta v = I_{sp} \cdot \ln\left[\frac{M_0}{M_0 - M_p}\right] \tag{2.5}$$

where

M_0 stands for the initial mass of the stage
M_p is the propellant mass of the stage
I_{sp} is the specific impulse.

To maximize the performance, one may maximize the $M_p : M_0$ ratio or increase the specific impulse I_{sp}. Much of the emphasis in propulsion technology has focused on enhancement of I_{sp} because of its major impact on the rocket Performance [136]. The I_{sp} is given by:

$$I_{sp} = \sqrt{\left[\frac{2\gamma}{(\gamma-1)}\right]\left(\frac{R_0}{M}\right)T_c\left[1-\left(\frac{p_e}{p_c}\right)^{\frac{(\gamma-1)}{\gamma}}\right]} \tag{2.6}$$

where,
γ is the ratio of specific heats
R is the universal gas constant
M is the molar mass of combustion products
T_c is the combustion temperature
P_e is the exit pressure at the nozzle,
P_c is the combustion pressure.

It is clear from Eq. (2.6) that the specific impulse depends on engine properties, that is, the ratio of exit and combustion pressure, and also on the properties of the combustion gases.

For a given engine design, the following properties of propellant will improve the specific impulse:

a. **High value of ratio of specific heats**. Polyatomic gases will have a higher value of specific heat at constant pressure (C_p) and hence have a high value of γ.
b. **Low molecular mass combustion products**.

c. **High combustion temperature.** Combustion temperature results from the energy release during combustion, that is, the heat of reaction and specific heat of combustion products. A higher combustion temperature is usually realized when the propellant ingredients have a positive heat of formation.
d. **High burn rate with acceptable pressure exponent (0.5–0.8).**
e. **High density.**

Let us understand how the above mentioned parameters work for energetic azido polymer based formulations to produce high performance propellants. The azido polymer based formulations are fuel rich and hence produce low molecular mass products during combustion. In addition, when azido polymers (which themselves have a positive heat of formation) are combined with energetic materials having positive heats of formation (RDX, HMX), higher combustion temperatures are possible. Both of these properties, combined with their inherent higher density and burn rates, are beneficial to obtaining high specific impulses from azido polymer formulations. However, the fuel-rich formulations require additional oxygen for efficient combustion, which otherwise leads to char formation, resulting in unstable combustion of the propellant. As oxygen has a high molar mass, this increases the mean molar mass of the combustion products. Hence the ratio $T_c : M$ (see Eq. (2.6)) has to be maximized to obtain a higher specific impulse. In practice, performance calculations are carried out using computational codes [137, 138] in order to predict which combination of oxidizer and binder provides the best performance.

Theoretical specific impulses for developmental formulations of RDX, CL-20, HNF, and ADN, combined with the azido polymer GAP are listed in Table 2.10. Aluminum (Al) has been added as a fuel to further improve the specific impulse of the formulations.

All the GAP formulations exhibit higher specific impulses as compared with the reduced smoke formulation based on the HTPB binder. The highest predicted specific impulses are for the aluminized GAP propellants of CL-20, ADN, and HNF.

Table 2.10 Theoretical specific impulse values of smokeless developmental formulations based on azido polymer GAP. From Refs. [138, 139].

Formulation	Specific impulse (s)	Flame temperature (K)
HTPB reduced smoke control	247	~1420
GAP/AN/Al	261.5	~2400
GAP/AN/CL-20/Al	263.7	–
GAP/CL-20/Al	273	–
GAP/ADN/Al	274.2	~3200
GAP/HNF/Al	272.6	~3300
GAP/TMETN/BTTN/RDX (60%)	243	~3000
GAP/TMETN/BTTN/CL-20 (60%)	252	–

The flame temperatures of GAP formulations are higher than those of the HTPB binder formulations. Even though the predicted performance parameters of the developmental formulations are promising, the sensitivity aspects of the oxidizers (especially the high impact sensitivity of HNF) should be taken into consideration when formulating safer propellants. Azido polymers provide definite performance advantages over inert polymers as binders in the propellant formulations.

2.10
Azido Polymers as Explosive Binders

Conventional PBX formulations contain binders of inert polymers. While the inert binders desensitize the hazardous explosive ingredients with which they are mixed, they also dilute or degrade the useful explosive energy obtainable from the explosive composite. When inert polymers are replaced by energetic polymers in the formulation, performance is enhanced due to the additional chemical energy provided by the energetic polymer. Moreover, as the energetic polymer content is increased, the loading level of the explosive filler could be correspondingly reduced to achieve a given performance requirement. The reduction in the amount of explosive filler in the energetic binder formulation is translated into a more favorable tradeoff between performance and safety properties than that which exists with conventional PBXs using inert polymers at present [140].

This section discusses the effectiveness of the azido polymer GAP as an explosive binder.

Radwan [141] measured the impact sensitivity of GAP/RDX and HTPB/RDX formulations using a Julius Peters impact sensitivity apparatus (Figure 2.26).

Both inert HTPB and energetic GAP binders are equally capable of improving the impact sensitivity of pure RDX. A formulation containing 15% GAP decreased the impact sensitivity of RDX by 100%, while that containing 21% GAP lowered the impact sensitivity by 211%. Addition of Al significantly decreases the impact sensitivity for both GAP and HTPB compositions, as Al is considered as an inert ingredient in these formulations. However, the GAP composition has much lower impact sensitivity compared with HTPB, for the same amount of RDX and Al.

The difference between the inert binder and energetic binder PBX compositions will become vivid when the performance parameters of the PBX compositions are compared. Velocity of detonation (VOD) is the important performance parameter of any explosive composition, which signifies the rate of propagation of a shock wave in an explosive. For a given explosive, VOD is directly proportional to its density [142]. As energetic binder compositions have higher densities, they are expected to provide higher VOD values than inert binder compositions. Figure 2.27 compares the experimental VOD values of RDX based PBX composition for varying contents of GAP and HTPB binders [141].

Clearly, addition of 12 wt% binder has reduced the VOD of RDX (8800 m/s [143]) by 14.5% for the GAP formulation and 19.3% for the HTPB. The reduction is due to

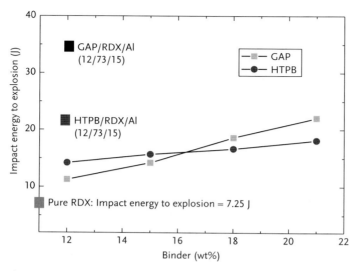

Figure 2.26 Comparison of impact sensitivities for RDX formulations with GAP and HTPB binders. Impact tests were carried out by using a 2 kg falling hammer. Values for the plot taken from Ref. [141].

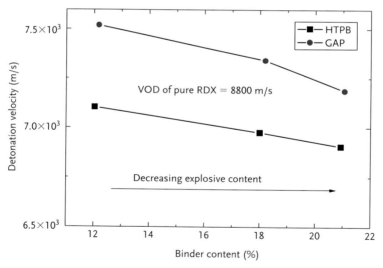

Figure 2.27 Variation of VOD with binder content for RDX based PBX composition. The binders used are HTPB and GAP. Values of the plot are taken from Ref. [141].

the replacement of high VOD RDX explosive with the binders in the formulation. In the GAP binder formulation, the reduction in the energetics of the PBX is compensated to an extent by the energetic nature of the GAP binder. On the other hand, the percentage reduction of VOD with respect to pure RDX is more for the HTPB binder formulation as the inert nature of the binder dilutes the overall energy of the formulation. Comparing the GAP and HTPB formulations, the GAP binder formulation delivers a higher performance for a given GAP loading level. Thus the energetic polymer GAP serves to achieve a more favorable tradeoff between performance and hazard properties as compared with the inert polymer formulations. This ability of the energetic polymer GAP has been utilized in PBX formulations intended for specific applications.

2.10.1
Azido Polymer Based PBX Formulations for Underwater Explosives

When an explosive charge is detonated in deep water, the total energy liberated in such an event is partitioned into radiated shock wave energy and bubble energy. The degree of partitioning depends on the nature of the explosive and the distance from the explosive event [144]. Shock wave energy and bubble energy are two critical parameters determining the underwater performance of an explosive. The shock wave shatters targets that are relatively closer to the detonation point and the bubble parameter damages surface targets, such as a destroyer [145].

Generally, the explosives used for underwater applications are aluminized explosives based on TNT (trinitrotoluene) and RDX. The reaction between aluminum with the primary detonation products and water liberates a huge amount of energy resulting in an increase in the bubble energy and in the shock wave energy. Composition H-6 is a widely used main charge filling for underwater blast weapons such as mines, depth charges, torpedoes, and mine disposal charges. It was first developed in the USA as an enhanced blast filling based on RDX/TNT/Al/wax. Even though RDX helps to reduce the sensitivity of TNT, the H-6 composition is regarded as too sensitive. Australia developed a PBX consisting of AP/bi-modal RDX/Al in a plasticized HTPB binder, known as PBXW-115, by completely eliminating TNT from the formulation [146]. The Australian formulated PBXW-115 was found to have better insensitive munition (IM) characteristics, higher shock wave, and relative bubble energies compared with the conventional H-6 and torpex explosives [147]. The PBXW-115 has been fully qualified as a possible insensitive war head fill by the US Navy and renamed as PBXN-111. ICT Germany [148] developed underwater formulations by replacing HTPB with the energetic binder GAP. Table 2.11 compares the bubble energies of the GAP based underwater explosives with those of PBXW-115 and H-6.

Table 2.11 Comparison of bubble energy values for various underwater explosives [147, 148].

Composition name (country)	Formulation	Bubble energy (MJ/kg)
GHX-85 (Germany)	RDX/Al/GAP (52/30/18)	3.49
GHX-99 (Germany)	RDX/Al/GAP (47/30/23)	3.56
PBXW-115 (Australia)	RDX/AP/Al/HTPB (20/43/25/12)	2.25
H-6 (Australia)	RDX/TNT/Al/Wax (45/30/20/5)	1.54

Replacing inert HTPB polymer with GAP improves the bubble energy of the formulations; thereby enhancing the performance of underwater explosives. Also, in the energetic polymer compositions, the improvement in bubble energy is achieved at a lower explosive content as compared with the inert binder compositions. This will render energetic binder compositions with better IM characteristics.

2.11
Tetrazole Polymers and Their Salts

Research interest in tetrazole polymers is generated by the high nitrogen content, considerable energetics, and the high thermal stability of the tetrazole heterocycle. Such polymers are considered to be prospective energetic binders for propellant and explosive formulations due to their high heats of formation. The tetrazole group is a mimic of the azido groups and releases comparable amounts of energy upon combustion. It is more stable and versatile than the azide group.

Poly(vinyl tetrazole) (PVT) is an important candidate among the tetrazole polymers [148] used in energetic material applications. The traditional method for the synthesis of PVT is the free-radical polymerization of the vinyl tetrazole monomer. This method has limitations due to the non-availability of the monomer [149]. Moreover, free-radical polymerization leads to materials of poorly controlled molecular weight distribution and chain-end functionalities. Alternatively, polymers with tetrazole pendant groups are prepared by chemical modification of polyacrylonitrile (PAN) (Scheme 2.13) [150]. The reaction is a representative of a group of 1,3-polar cycloadditions, collectively termed as "click chemistry" reactions, which are characterized by simple reaction conditions and higher yields [151].

The energetics of the polymer are determined by the vinyl tetrazole content, which increases with the molecular weight of the PAN used. For example, a tetrazole content of 97.5% could be achieved by using PAN of molecular weight 200 000. PVT with the highest tetrazole content shows an explosive thermal

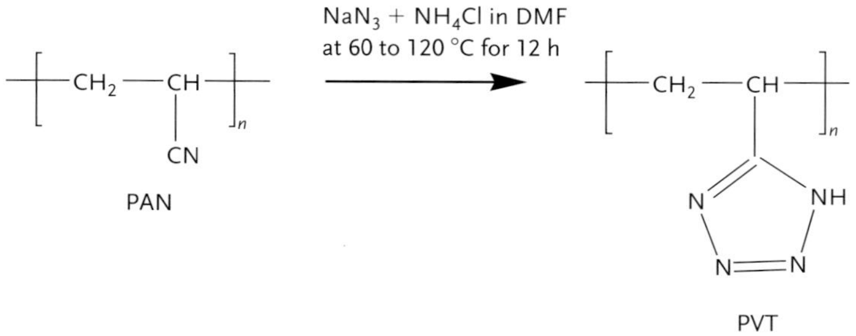

Scheme 2.13 Synthetic scheme for the preparation of poly(vinyl tetrazole) from polyacrylonitrile [150].

degradation together with the release of a large amount of heat. However, owing to their high viscosity, processability of energetic formulations with high molecular weight PVT polymers will be energy intensive. Hence, Chafin et al. [152] synthesized low molecular weight multifunctional hydroxyethyl tetrazoles for energetic binder applications (Scheme 2.14). The synthetic procedure involves transforming nitrile into tetrazole using sodium azide and zinc salt in water [153, 154], followed by alkylation with 2-chloroethanol to give a mixture of 1- and 2-alkylated products, which were not separated. The diols synthesized by Scheme 2.14 could be further cross-linked by the isocyanate curing agents to form energetic binders.

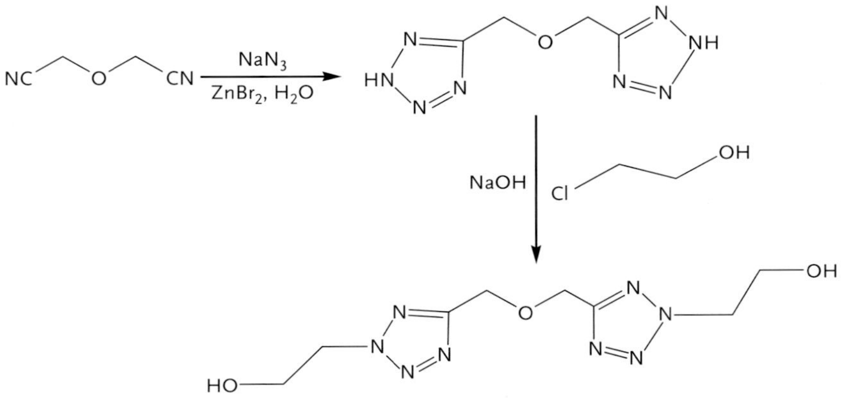

Scheme 2.14 Synthetic scheme for di-functional hydroxyethyl tetrazole [153].

Similarly, trifunctional and tetrafunctional tetrazoles, which find application as high nitrogen content cross-linking materials, were also prepared in high yields by the above synthetic procedure by changing the starting nitrile compounds [153].

Vinyl tetrazole polymers are glassy and somewhat brittle (they possess high glass transition temperatures). An ideal energetic binder should be tough, soft, dense, and capable of dissolving large proportions of plasticizers without exudation. The tetrazole polymer poly(2-methyl-5-vinyl tetrazole) (PMVT) has been known for over 25 years. Among the tetrazole polymers, alkylated vinyl tetrazoles have evoked considerable interest because of their enhanced thermal stability. The high thermal stability is ascribed to the low frequency twisting vibrations of the alkyl groups, which facilitate rapid dissipation of thermal energy stored in the heterocycle [155]. In order to make PMVT suitable as an energetic binder with the required properties, it is mixed with polyethylene glycol (PEG) in equimolar proportions to form a compatible blend with the desirable characteristics [156]. The structure of PMVT/PEG polymer blend is shown in Figure 2.28.

The two polymers form interpenetrating network strengthened by hydrogen bonded interactions between the methine hydrogen of the PMVT and the ether

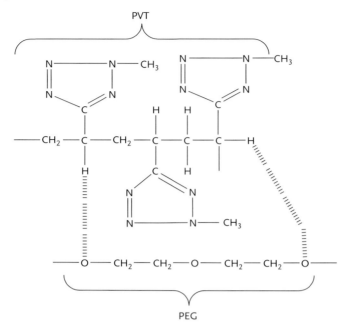

Figure 2.28 Structure of PMVT-PEG blend [156].

oxygen of the PEG, resulting in a fully amorphous blend with a single glass transition temperature (the cured blend has a T_g of $-4\,°C$). Blend formation also results in a decrease in the free volume and a consequent increase in density and thus specific energy. This polymer blend is found to be a desensitizing energetic binder for PBX formulations. A PBX composition containing 91% by weight RDX and 9% by weight of PMVT/PEG has a better drop-weight sensitivity ($H_{50} = 1.32$ m) than composition B ($H_{50} = 0.5$–0.8 m), comprising of a mixture of TNT and RDX. It is suggested that [157] the pendant tetrazole groups act as energy sinks and absorb energy by delocalization into its many π-orbitals. In addition to the delocalization, extra energy dissipation arises from the vibrational relaxations of the tetrazole and methyl groups.

Energetic salts of nitrogen heterocycles (such as tetrazoles and triazoles), known generally as ionic liquids, have gained considerable interest as "green" energetic materials as these salts generate environmentally acceptable nitrogen as the end product of an explosion or propulsion. Compared with the atomically similar non-ionic analogues, the energetic ionic liquids possess attractive properties, namely higher densities, low vapor pressure (hence no vapor toxicity), and low melting-points [158, 159]. The low vapor pressure feature means they are categorized as "green solvents."

In the scientific literature, polymer salts obtained from nitrogen heterocycles have been widely studied for their application as highly conducting ion electrolytes and absorbers for carbon dioxide [160]. However, the utility of these salts as energetic materials has rarely been exploited. Preparation of a triaminoguanidinium salt of 5-vinyl tetrazole polymers was disclosed as early as 1968 [161] as a possible binder ingredient for solid and hybrid rocket fuel formulations. This energetic polymer salt is prepared by the reaction of one equivalent of the free base, triaminoguanidine, with 5-vinyl tetrazole polymer in an inert atmosphere at temperatures ranging from -20 to $50\,°C$. The salt was observed to have a heat of formation of 364 kJ/mol and shows no measurable impact or electrical sensitivity ($H_{50} = 1.40$ m (2 kg hammer), ESD $= 6.56$ J, auto-ignition $= 300\,°C$).

In recent work Shreeve and coworkers synthesized energetic polymer salts from 1-vinyl-1,2,4-triazole derivatives [162]. They were synthesized either by the free-radical polymerization of 1-vinyl-1,2,4-triazolium salts or by protonation of poly(1-vinyl-1,2,4-tetrazole) using organic or inorganic acids. Compared with the monomer salts, the polymer salts have higher densities (1.5 g/cm^3) and good thermal stability.

Kizhnyaev et al. [163] utilized the gelling property of alkali or ammonium salts of PVT in the presence of transition metal ions for the design of water filled explosive composites. The sodium salt of PVT forms an energetic gel in the presence of chromium (Cr^{3+}) ion, functioning as an efficient energetic fuel binder for water filled explosive composites consisting of ammonium nitrate oxidizer.

2.12
N–N-Bonded Epoxy Binders

Polymers with epoxy groups find application as propellant binders because of the good combustion characteristics, low viscosity, superior mechanical strength, toughness, and unique adhesive properties [164]. Energetic polymers comprising of N–N bonds in the backbone of the polymer binder and having epoxy terminal groups that are cured easily were developed as binders for hybrid and solid rocket propellants [165]. The epoxy binders have been synthesized by reacting N,N'-aliphatic dicarboxyl-bis-hydrazones based on malonic, adipic, and sebacic dihydrazides with epichlorohydrin (Scheme 2.15) [166].

The resins could be cured by the epoxy curing agent 4,4′-diaminodiphenylmethane (DDM). Thermally, all the resins exhibited an exothermic decomposition peak at 220 °C, which does not vary with the number of methylene spacer groups. Exothermic decomposition is due to the cleavage of the N–N hydrazine bond in the polymer [167]. The glass transition temperatures of the cured resins vary from -13 to $55\,°C$ [168] and are the lowest T_g recorded for the resin with the largest number of methylene spacer groups. Heat of combustion values of the binder increase with the number of methylene spacer groups: 6.35 kcal/g for the malonyl substitute to 7.13 kcal/g for the sebacoyl substitute [166]. Thus, the glass transition temperatures and heat of combustion values could be tailored by suitably changing the macromolecular architecture. They exhibited superior ignition and burn rates over the conventional butadiene binders [169]. Furthermore, energetic

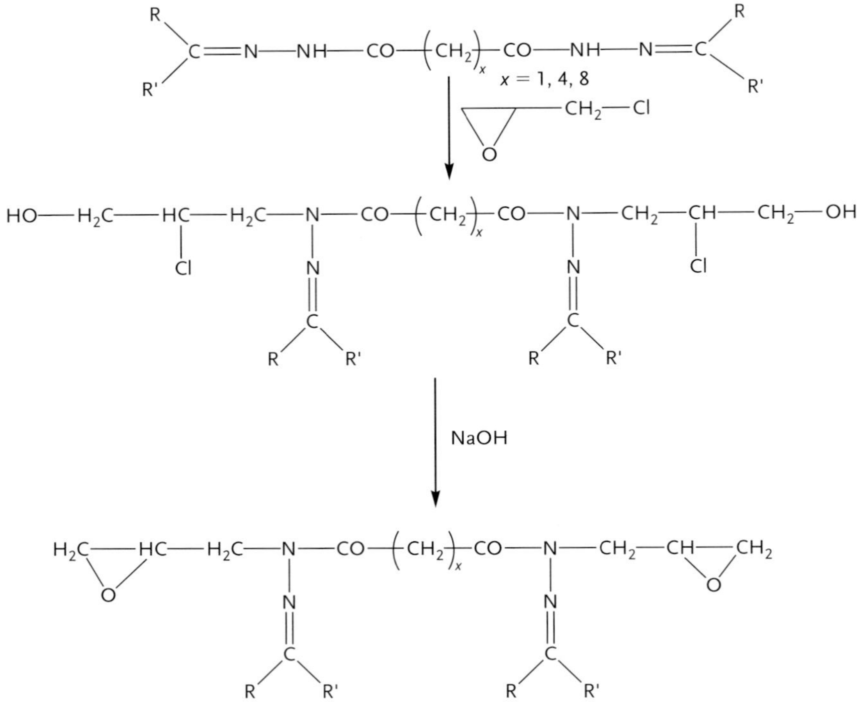

Scheme 2.15 Preparative scheme for N–N bonded epoxy resins [166].

composites of these resins with an AP oxidizer exhibited a wider range of mechanical properties adequate for propellant binders, with varying numbers of methylene spacer groups and natures of substituent groups in the resin [166].

References

1. Badgujar, D.M., Talwar, M.B., Asthana, S.N., and Mahulikar, P.P. (2008) Advances in science and technology of modern energetic materials: an overview. *J. Hazard. Mater.*, **151**, 289–305.
2. Earl, R. (1984) Use of polymeric ethylene oxides in the preparation of GAP. US Patent 4,486,351.
3. Ahad, E. (1987) Process for preparation of hydroxy terminated polyethers. European Patent 87310476.4.
4. Frankel, M.B. and Flanagan, J.E. (1981) Energetic hydroxy terminated azide polymer. US Patent 4,268,450.
5. Frankel, M.B., Witucki, E.F., and Woolery, D.O. (1983) Aqueous process for quantitative conversion of PECH to GAP. US Patent 4,379,894.
6. Wagner, R.I. (1991) Glycidyl azide polymer synthesis by molten method. US Patent 5,055,600.
7. Ampleman, G. (1990) Synthesis of a new class of glycidyl azide polymers. US Patent 5,256,804.
8. Ahad, E. (1990) Branched hydroxy terminated aliphatic polyethers. US Patent 4,882,395.
9. Labrecque, B. and Roy, A. (1990) Pilot plant study of an energetic azide

polymer synthesis, TTCP, W4. Propulsion Technology, 15th Meeting, Valcartier, Canada, DREV.

10 Barna, J.A., Groenewed, P.G., Holden, H.W., and Leonard, J.A. (1994) Pilot plant for azido polymers: Branched GAP process and polymers. International Symposium on Energetic Materials Technology, Orlando, Florida, American Defense Preparedness Association.

11 Desai, H. (1996) Telechelic polyoxetanes, in *Polymeric Materials Encyclopedia*, Vol. **11** (ed. J.C. Salamone), CRC Press, Boca Raton, pp. 8268–8279.

12 Cheradame, H., Andreolety, J.-P., and Rousset, E. (1991) Synthesis of polymers containing pseudohalide groups by cationic polymerization. Part 1: homopolymerization of 3,3-bis (azidomethyl) oxetane and its copolymerization with 3-chloromethyl-3-(2,5,8 -trioxadecyl) oxetane. *Makromol. Chem.*, **192**, 901–918.

13 Murphy, E.A., Ntozakhe, T., Murphy, C.J., Fay, J.J., Sperling, L.H., and Manser, G.E. (1989) Characterization of poly (BEMO) and poly (BAMO) and their block copolymers. *J. Appl. Polym. Sci.*, 267–281.

14 Earl, R.A. and Elmslie, J.S. (1981) Preparation of hydroxy-terminated poly (BAMO). US Patent 4,405,762.

15 Barbieri, U., Polacco, G., and Massimi, R. (2006) Synthesis of energetic polyethers from halogenated precursors. *Macromol. Symp.*, **234** (1), 51–58.

16 Stockel, R.F. and Valenti, P.C. (1977) Method of preparing 3,3-bis (chloromethyl) oxetane. US Patent 4,031,110.

17 Hardenstine, K.E., Henderson, G.V.S., Sperling, L.H., Murphy, G.E., and Manser, C.J. (1985) Crystallization behavior of poly (BEMO) and poly (BAMO). *J. Polym. Sci. Polym. Phys. Ed.*, **23**, 1597–1609.

18 Jutier, J.-J., De Guzbourg, A., and Prud'Homme, R.E. (1999) Synthesis and characterization poly (BAMO-*co*-ε caprolactone)s. *J. Polym. Sci. A: Polym. Chem.*, **37**, 1027–1039.

19 Hsiue, G.-H., Liu, Y.-L., and Chiu, Y.-S. (1993) Tetrahydrofuran and BCMO triblock copolymers synthesized by two-end living cationic polymerization. *J. Polym. Sci. A: Polym. Chem.*, **31**, 3371–3375.

20 Manser, G.E. (1993) Thermoplastic elastomers having alternate crystalline structure for use as high energy binders. US Patent 5,210,153.

21 Liu, Y.-L., Hsiue, G.-H., and Chiu, Y.-S. (1995) Studies on polymerization mechanism of 3-nitratomethyl-3'-methyloxetane and 3-azidomethyl-3'-methyloxetane and synthesis of their respective copolymers with tetrahydrofuran. *J. Polym. Sci. A: Polym. Chem.*, **33**, 1697–1613.

22 Cheun, Y.-G., Kim, J.-S., and Jo, B.-Y. (1994) A study on cationic polymerization of energetic oxetane derivatives. Proceedings of 25th International Annual Conference of ICT, Karlsruhe, 71/1–71/14.

23 Talukder, M.A.H. (1991) Ring opening polymerization of oxetanes by cationic initiators. Polymerization of 3-azidomethyl 3-methyl oxetane with bis(chlorodimethylsilyl)-benzene/silver hexafluoroantimonate initiating system. *Makromol. Chem. Macromol. Symp.*, **42/43**, 501–511.

24 Min, B.S. (2008) Characterization of the plasticized GAP/PEG and GAP/PCL block copolyurethane binder matrices and its propellants. *Propellants Explos. Pyrotech.*, **33** (2), 131–138.

25 Randall, D. and Lee, S. (2002) *The Polyurethanes Book*, John Wiley & Sons, Ltd, Chichester.

26 Hepburn, C. (1982) *Polyurethane Elastomers*, Applied Science Publishers Ltd, London and New York.

27 Kezkin, S. and Ozkar, S. (2001) Kinetics of polyurethane formation between glycidyl azide polymer and a triisocyanate. *J. Appl. Polym. Sci.*, **81**, 918–923.

28 Reshmi, S., Sadhana, R., Varghese, T.L., Ravindran, P.G., Kannan, K.G., and Ninan, K.N. (2004) Pre-gel cure kinetics of glycidyl azide polymer with different diisocyanates. Proceedings of

35th International Annual Conference of ICT, 99/1–99/13.
29 Kothandaraman, H., and Sultan Nasar, A. (1993) The kinetics of polymerization reaction of toluene diisocyanate with HTPB polymer. *J. Appl. Polym. Sci.*, **50**, 1611–1617.
30 Kincal, D. and Ozkar, S. (1997) Kinetics study of the reaction between hydroxyl-terminated polybutadiene and isophorone diisocyanate in bulk by quantitative FTIR spectroscopy. *J. Appl. Polym. Sci.*, **66**, 1979–1983.
31 Yee, R.Y. and Adicoff, A. (1976) Polymerization kinetics in the propellants of hydroxyl-terminated polybutadiene – isophorone diisocyanate system. *J. Appl. Polym. Sci.*, **20**, 1117–1124.
32 Aranguren, M.I. and Williams, R.J.J. (1986) Kinetics and statistical aspects of polyurethane formation from toluene diisocyanate. *Polymer*, **27**, 425–430.
33 Han, J.L., Yu, C.H., Lin, Y.H., and Hsieh, K.H. (2008) Kinetic studies of urethane and urea reactions of isophorone diisocyanate. *J. Appl. Polym. Sci.*, **107**, 3891–3902.
34 Houghton, R.P., and Mulvaney, A.W. (1996) Mechanism of tin (IV)-catalysed urethane formation. *J. Organomet. Chem.*, **518**, 21–27.
35 Reed, R. and Chan, M.L. (1983) Propellant binders cure catalyst. US Patent 4,379,903.
36 Lipshitz, S.D. and Macosko, C.W. (1976) Rheological changes during a urethane network formation. *Polym. Eng. Sci.*, **16** (12), 803–810.
37 Korah Bina, C., Kannan, K.G., and Ninan, K.N. (2004) DSC study on the effect of isocyanates and catalysts on HTPB cure reaction. *J. Therm. Anal. Calorim.*, **78**, 753–760.
38 Kasicki, H., Pekel, F., and Ozkar, S. (2001) Curing characteristics of glycidyl azide polymer based binders. *J. Appl. Polym. Sci.*, **80**, 65–70.
39 Eroglu, M.S. and Gluven, O. (1998) Characterization of network structure of poly (glycidyl azide) elastomers by swelling, solubility and mechanical measurements. *Polymer*, **39** (5), 1173–1176.
40 Eroglu, M.S. (1998) Characterization of network structure of HTPB elastomers prepared by different reactive systems. *J. Appl. Polym. Sci.*, **70**, 1129–1135.
41 Mathew, S., Manu, S.K., and Varghese, T.L. (2008) Thermomechanical and morphological characteristics of cross-linked GAP and GAP-HTPB networks with different diisocyanates. *Propellants Explos. Pyrotech.*, **33** (2), 146–152.
42 Sekkar, V., Narayanaswamy, K., Scariah, K.J., Nair, P.R., and Sastri, K.S. (2006) Studies on urethane-allophanate networks based on HTPB: modeling of network parameters and correlation to mechanical properties. *J. Appl. Polym. Sci.*, **101**, 2986–2994.
43 Haska, S.B., Bayramli, E., Pekel, F., and Ozkar, S. (1997) Mechanical properties of HTPB-IPDI based elastomers. *J. Appl. Polym. Sci.*, **64** (12), 2347–2354.
44 Wingborg, N. (2001) Increasing the tensile strength of HTPB with different isocyanates and chain extenders. *Polym. Test.*, **21**, 283–287.
45 Manzara, A.P. (1997) Azido polymers having improved burn rates. US Patent 5,681,904.
46 Keicher, T., Kuglstatter, W., Eisele, S., Wetzel, T., Kaiser, M., and Krause, H. (2010) Isocyanate-free curing of glycidyl-azide polymer. 41st International Conference of ICT, 12/1–12/15.
47 Keicher, T., Kuglstatter, W., Eisele, S., Wetzel, T., Kaiser, M., and Krause, H. (2007) Proceedings of 39th Annual Conference of ICT, 66/1–66/13.
48 Russell, R. Jr. (2000) Triazole cross-linked polymers. US Patent 6,103,029.
49 Benson, F.R. and Savell, W.L. (1950) The chemistry of the vicinal triazoles. *Chem. Rev.*, **46**, 1–68.
50 Katiritzky, A.R. and Singh, S.K. (2002) Synthesis of C-carbamoyl-1,2,3-triazoles by microwave-induced 1,3-dipolar cycloaddition of organic azides to acetylenic amides. *J. Org. Chem.*, **67** (25), 9077–9079.
51 Katritzky, A.R., Meher, N.K., Hanci, S., Gayanda, R., Tala, S.R., Mathai, S., Duran, R.S., Bernard, S., Sabri, F., Singh, S.K., Doskocz, J., and

Ciaramitaro, D.A. (2008) Preparation and characterization of 1,2,3-triazole-cured polymers from endcapped azides and alkynes. *J. Polym. Sci. A: Polym. Chem.*, **46**, 238–256.

52 Ciaramitaro, D. (2005) Triazole cross-linked polymers in recyclable energetic compositions and method of preparing the same. US Patent 6,872,266 B1.

53 Chen, J.K. and Brill, T.B. (1991) Thermal decomposition of energetic materials 54. Kinetics and near-surface products of azide polymers AMMO, BAMO and GAP in simulated combustion. *Combust. Flame*, **87**, 57–168.

54 Oyumi, Y. and Brill, T.B. (1986) Thermal decomposition of energetic materials 12. Infrared spectral and rapid thermolysis studies of azide-containing monomers and polymers. *Combust. Flame*, **65**, 127–135.

55 Arisawa, H. and Brill, T.B. (1998) Thermal decomposition of energetic materials 71. Structure-decomposition and kinetic relationships in flash pyrolysis of glycidyl azide polymer (GAP). *Combust. Flame*, **112**, 533–544.

56 Selim, K., Ozkar, S., and Yilmaz, L. (2000) Thermal characterization of glycidyl azide polymer (GAP) and GAP based binders for composite propellants. *J. Appl. Polym. Sci.*, **77**, 538–546.

57 Pfeil, A. and Lobbecke, S. (1997) Controlled pyrolysis of the new energetic binder azide polyester PAP-G. *Propellants Explos. Pyrotech.*, **22**, 137–142.

58 (a) Kuibida, L.V., Korobeinichev, O.P., Shmakov, A.G., Volkov, E.N., and Paletsky, A.A. (2001) Mass spectrometric study of combustion of GAP and ADN-based propellants. *Combust. Flame*, **126**, 1655–1661; (b) Korobeinichev, O.P., Kuibida, L.V., Volkov, E.N., and Shmakov, A.G. (2002) Mass spectrometric study of combustion and thermal decomposition of GAP. *Combust. Flame*, **129**, 136–150.

59 Fazlioglu, H. and Hacaloglu, J. (2002) Thermal decomposition of glycidyl azide polymer by direct insertion probe mass spectrometry. *J. Anal. Appl. Pyrol.*, **63**, 327–338.

60 Wang, T., Li, S., Yang, B., Huang, C., and Li, Y. (2007) Thermal decompositon of glycidyl azide polymer studied by synchrotron photoionization mass spectrometry. *J. Phys. Chem. B*, **111**, 2449–2455.

61 Kubota, N. and Sonobe, T. (1988) Combustion mechanism of the azide polymer. *Propellants Explos. Pyrotech.*, **13**, 172–177.

62 Haas, Y., Eliahu, Y.B., and Welner, S. (1994) Infrared laser-decomposition of GAP. *Combust. Flame*, **96**, 212–220.

63 Faber, M., Harris, S.P., and Srivastava, R.D. (1984) Mass spectrometric kinetic studies on several azido polymers. *Combust. Flame*, **55**, 203–211.

64 Ringuette, S., Stowe, R., Dubois, C., Charlet, G., Kwok, Q., and Jones, D.E. G. (2006) Deuterium effect on thermal decomposition of deuterated GAP: 1. Slow thermal analysis with a TGA-DTA-FTIR-MS. *J. Energetic Mater.*, **24**, 307–320.

65 Puduppakkam, K.V. and Beckstead, M.W. (2005) Combustion modelling of glycidyl azide polymer with detailed kinetics. *Combust. Sci. Tech.*, **177**, 1661–1691.

66 Eroglu, M.S. and Guven, O. (1996) Thermal decomposition of poly (glycidyl) azide as studied by high temperature FTIR and thermogravimetry. *J. Appl. Polym. Sci.*, **61** (2), 201–206.

67 Panda, S.P., Sahu, S.K., Sadafule, D.S., and Thakur, J.V. (2000) Role of curing agents on decomposition and explosion of glycidyl azide polymers. *J. Propul.*, **16** (4), 723–725.

68 Brill, T.B., Brush, P.J., James, K.J., Shepherd, J.E., and Pfeiffer, K.J. (1992) T-Jump/FTIR spectroscopy: a new entry into the rapid pyrolysis chemistry of solids and liquids. *Appl. Spectrosc.*, **46** (6), 900–911.

69 Lee, Y.J., Tang, C.J., Kudva, G., and Litzinger, T.A. (1998) Thermal decompositon of 3,3′-azidomethyl-oxetane. *J. Propul. Power*, **14** (1), 37–44.

70 Golub, M.A. and Gargiulo, R.J. (1972) Thermal degradation of 1,4-polyisoperene and 1,4-polybutadiene. *J. Polym. Sci. B: Polym. Lett.*, **10** (1), 41–49.

71 Brazier, D.W., and Schwartz, N.V. (1978) The effect of heating rate on thermal degradation of polybutadiene. *J. Appl. Polym. Sci.*, **22**, 113–124.

72 McCreedy, K. and Keskkula, H. (1979) Effect of thermal crosslinking on decomposition of polybutadiene. *Polymer*, **20**, 1155–1159.

73 Lu, Y.C. and Kuo, K.K. (1996) Thermal decomposition study of HTPB solid fuel. *Thermochim. Acta*, **275**, 181–191.

74 Beck, W.H. (1987) Pyrolysis studies of polymer materials used as binders in composite propellants: a review. *Combust. Flame*, **70**, 171–190.

75 Chen, J.K. and Brill, T.B. (1991) Chemistry and kinetics of HTPB and diisocyanate–HTPB polymers during slow decomposition and combustion like conditions. *Combust. Flame*, **87**, 217–232.

76 Arisawa, H. and Brill, T.B. (1996) Flash pyrolysis of HTPB II: implications of the kinetics to combustion of organic polymers. *Combust. Flame*, **106**, 144–154.

77 Vyazovkin, S. and Wight, C.A. (1997) Kinetics in solids. *Annu. Rev. Phys. Chem.*, **48**, 125–149.

78 Vyazovkin, S. (2004) Thermal analysis. *Anal. Chem.*, **74**, 3299–3312.

79 Oyumi, Y. (1992) Thermal decomposition of azide polymers. *Propellants Explos. Pyrotech.*, **17**, 226–231.

80 Raemaekers, K.G.H. and Bart, J.C.J. (1997) Application of simultaneous thermogravimetry-mass spectrometry in polymer analysis. *Thermochim. Acta*, **295**, 1–58.

81 Long, G.T., Vyazovkin, S., Gamble, N., and Wight, C.A. (2001) Hard to swallow dry: kinetics and mechanism of anhydrous thermal decomposition of acetylsalicylic acid. *J. Pharm. Sci.*, **91** (3), 800–809.

82 Pisharath, S. and Ang, H.G. (2007) Synthesis and thermal decomposition of GAP-Poly (BAMO) copolymer. *Polym. Degrad. Stab.*, **92**, 1365–1377.

83 Jones, D.E.G., Malechaux, L., and Augsten, R.A. (1994) Thermal analysis of GAP TRIOL, an energetic azide polymer. *Thermochim. Acta*, **242**, 187–197.

84 Feng, H.T., Mintz, K.J., Augsten, R.A., and Jones, D.E.G. (1998) Thermal analysis of branched GAP. *Thermochim. Acta*, **311**, 105–111.

85 Du, T. (1989) Thermal decomposition studies of solid propellant binder HTPB. *Thermochim. Acta*, **138**, 189–197.

86 Kubota, N. and Sonobe, T. (1988) Combustion mechanism of azide polymer. *Propellants Explos. Pyrotech.*, **13**, 172–177.

87 Hori, K. and Kimura, M. (1996) Combustion mechanism of glycidyl azide polymer. *Propellants Explos. Pyrotech.*, **21**, 160–165.

88 Miyazaki, T. and Kubota, N. (1992) Energetics of BAMO. *Propellants Explos. Pyrotech.*, **17**, 5–9.

89 Bazaki, H. and Kubota, N. (1991) Energetics of AMMO. *Propellants Explos. Pyrotech.*, **16**, 68–72.

90 Kubota, N. (1995) Combustion of energetic azide polymers. *J. Propul. Power*, **11**, 677–682.

91 Davidson, J.E. and Beckstead, M.W. (1997) A mechanism and model for GAP combustion. Paper 97-0592. AIAA Aerospace Meetings and Exhibit, Reno, NV.

92 Kubota, N. (1984) Survey of rocket propellants and their combustion characteristics, in *Fundamentals of Solid Propellant Combustion, Progress in Astronautics and Aeronautics*, Vol. **90** (eds K.K. Kuo and M. Summerfield), AIAA, New York, pp. 1–52.

93 Brill, T.B., Brush, P.J., Gray, P., and Kinloch, S.A. (1992) Condensed phase chemistry of explosives and propellants at high temperature: HMX, RDX and BAMO. *Philos. Trans. Phys. Sci. Eng.*, **339**, 377–385.

94 Beckstead, M.W. (2000) Overview of combustion mechanisms and flame structures for advanced solid

propellants, in *Solid Propellant Chemistry, Combustion and Motor Interior Ballistics, Progress in Astronautics and Aeronautics*, Vol. **185** (eds V. Yang, T.B. Brill, and W.-Z. Ren), AIAA, New York, pp. 267–285.

95 Sinditskii, V., Fogelzang, A.E., Egorshev, V.Yu., Serushkin, V.V., and Kolesov, V.I. (2000) Effect of molecular structure on combustion of polynitrogen energetic materials, in *Solid Propellant Chemistry, Combustion and Motor Interior Ballistics, Progress in Astronautics and Aeronautics*, Vol. **185** (eds V. Yang, T.B. Brill, and W.-Z. Ren), AIAA, New York, pp. 99–128.

96 Kim, E.S., Yang, V., and Liau, Y.-C. (2002) Modeling of HMX/GAP pseudo-propellant combustion. *Combust. Flame*, **131**, 227–245.

97 (a) Zeman, S. (2002) Modified Evans–Polanyi–Semenov relationship in the study of chemical micromechanism governing detonation and initiation of individual energetic materials. *Thermochim. Acta*, **384**, 137–154; (b) Zeman, S. (2006) New aspects of initiation reactivities of energetic materials demonstrated on nitramines. *J. Hazard. Mater.*, **132**, 155–164.

98 Oyumi, Y., Inokami, K., Yamazaki, K., and Matsumoto, K. (1993) Thermal decomposition of BAMO/HMX propellants. *Propellants Explos. Pyrotech.*, **18**, 62–68.

99 Ger, M.D., Hwu, W.H., and Huang, C.C. (1993) A study on the thermal decomposition of mixtures containing an energetic binder and a nitramine. *Thermochim. Acta*, **224**, 127–140.

100 (a) Bohn, M.A., Hammerl, A., Harris, K., and Klapotke, T.M. (2005) The elimination of NO_2 from mixtures of nitramines HMX, RDX and CL20 with the energetic binder glycidyl azide polymer – a computational study I. Central European. *J. Energetic Mater.*, **2**, 29–44; (b) Bohn, M.A., Hammerl, A., Harris, K., and Klapotke, T.M. (2005) Further decomposition pathways of mixtures of nitramines HMX, RDX and CL20 with the energetic binder glycidyl azide polymer – a computational study II. Central European. *J. Energetic Mater.*, **2**, 3–19; (c) Bohn, M.A., Hammerl, A., Harris, K., and Klapotke, T.M. (2005) Interactions between the nitramines RDX, HMX and CL20 with the energetic binder GAP. Proceedings of 8th Seminar on New Trends in Research of Energetic Materials, University of Pardubice, 490–497.

101 Oxley, J.C., Kooh, A.B., Szekeres, R., Zheng, W. (1994) Mechansim of nitramine thermolysis. *J. Phy. Chem.*, **98**, 7004–7008.

102 Karaghiosoff, K., Klapotke, T.M., Michailovski, A., and Holl, G. (2002) 4,10-dinitro-2,6,8,12-tetraoxa-4,10-diazaisowurtzitane. *Acta Crystallogr.*, **C58**, 580–581.

103 Pisharath, S. and Ang, H.-G. (2007) Thermal decomposition kinetics of a mixture of energetic polymer and nitramine oxidizer. *Thermochim. Acta*, **459**, 26–33.

104 Beckstead, M.W., Puduppakkam, K., Thakre, P., and Yang, V. Modeling of combustion and ignition of solid propellant ingredients. *Prog. Energetic Combust. Sci.*, **33**, 497–551 (2007).

105 Litzinger, T.A. Lee, Y.J., and Tang, C.-J. (2000) Experimental studies of nitramine/azide propellant combustion, in *Solid Propellant Chemistry, Combustion and Motor Interior Ballistics, Progress in Astronautics and Aeronautics*, Vol. **185** (eds V. Yang, T.B. Brill, and W.-Z. Ren), AIAA, New York, pp. 355–379.

106 Kubota, N. and Kuwahara, T. (1997) Combustion of GAP/HMX and GAP/TAGN energetic composite. *Propellants Explos. Pyrotech.*, **25**, 86–91.

107 Liau, Y.-C., Yang, V., and Thynell, S.T. (2000) Modeling of RDX/GAP propellant combustion with detailed chemical kinetics, in *Solid Propellant Chemistry, Combustion and Motor Interior Ballistics, Progress in Astronautics and Aeronautics*, Vol. **185** (eds V. Yang, T.B. Brill, and W.-Z. Ren) AIAA, New York, pp. 477–500.

108 Paletsky, A.A., Volkov, E.N., Korobeinichev, O.P., and

Tereshchenko, A.G. (2007) Flame structure of composite pseudo-propellants based on nitramines and azide polymers at high pressure. *Proc. Combust. Inst.*, **31**, 2079–2087.

109 Kubota, N. and Sonobe, T. (1990) Burning rate catalysis of azide/nitramine propellant. Proceedings of 23rd International Symposium on Combustion, The Combustion Institute, Pittsburgh, 1331–1337.

110 Yang, R., Thakre, P., Liau, Y.-C., and Yang, V. (2006) Formation of dark zone and temperature plateau in solid propellant flames. *Combust. Flame*, **145**, 38–58.

111 Lengelle, G., Duterque, J., and Trubert, J.F. (2000) Physico-chemical mechanisms of solid propellant combustion, in *Solid Propellant Chemistry, Combustion and Motor Interior Ballistics, Progress in Astronautics and Aeronautics*, Vol. **185** (eds V. Yang, T.B. Brill, and W.-Z. Ren), AIAA, New York, pp. 287–334.

112 Yoon, J.-K., Thakre, P., and Yang, V. (2006) Modeling of RDX/GAP/BTTN pseudo-propellant combustion. *Combust. Flame*, **145**, 300–315.

113 Parr, T.P. and Hanson-Parr, D. (2001) RDX/GAP/BTTN propellant. *Flame Studies*, **127**, 1895–1905.

114 Li, J. and Litzinger, T.A. (2007) Laser-driven decomposition and combustion of BTTN/GAP. *J. Propul. Power*, **23**, 166–174.

115 Roos, B.D. and Brill, T.B. (2001) Thermal decomposition of energetic materials 81. Flash pyrolysis of GAP/RDX/BTTN propellant combinations. *Propellants Explos. Pyrotech.*, **26**, 213–220.

116 Parr, T. and Hanson Parr, D. (2004) Cyclotetramethylene tetranitramine/GAP/BTTN propellant flame structure. *Combust. Flame*, **137**, 38–49.

117 Golfier, M., Graindorge, H., Longevaille, Y., and Mace, H. (1998) New energetic materials and their applications in energetic materials. Proceedings of 29th International Conference of ICT, Karlsruhe, 3/1-3/17.

118 Oommen, C. and Jain, S.R. (1999) Ammonium nitrate: a promising rocket propellant oxidizer. *J. Hazard. Mater.*, **A67**, 253–281.

119 Nakajima, C., Saito, T., Yamaya, T., and Shimoda, M. (1998) The effect of chromium compounds on PVA-coated AN and GAP binder pyrolysis and PVA-coated AN/GAP combustion. *Fuel*, **77**, 321–326.

120 Oyumi, Y., Kimura, E., Hayakawa, S., Nakashita, G., and Kato, K. (1996) Insensitive munitions and combustion characteristics of GAP/AN composite propellants. *Propellants Explos. Pyrotech.*, **21**, 271–275.

121 Menke, K., Bohnlein-Mauβ, J., and Schubert, H. (1996) Characteristic properties of AN/GAP propellants. *Propellants Explos. Pyrotech.*, **21**, 139–145.

122 Judge, M.D. and Lessard, P. (1996) An advanced GAP/AN/TAGN propellant. *Propellants Explos. Pyrotech.*, **32**, 175–181.

123 Schoyer, H.F.R., Schnorhk, A.J., Mul, J.M., Gadiot, G.M.H.J.L., and Meulenbrugge, J.J. (1995) High performance propellants based on hydrazinium nitroformate. *J. Propul. Power*, **11**, 856–869.

124 Schoyer, H.F.R., Welland–Veltmans, W.H.M., Louwers, J., Korting, P.A.O.G., van der Heijden, A.E.D.M., Keizers, H.L.J., and van den Berg, R.P. (2002) Overview of the development of hydrazinium nitroformate based propellants. *J. Propul. Power*, **18**, 138–145.

125 Zevenbergen, J.F., Pekalski, A.A., van der Heijden, A.E.D.M., Keizers, H.L.J., van den Berg, R.P., Welland–Veltmans, W.H.M., and Wierckx, F.J.M. (2005) Propulsion and energetic materials research in Netherlands. AIAA Paper 2005-3848. 41st AIAA/ASME/SAE/ASEE Joint Propulsion Conference & Exhibit, Tucson, Arizona, USA.

126 Williams, G.K. and Brill, T.B. (1995) Thermal decomposition of energetic materials 67. Hydrazinium nitroformate (HNF) rates and pathways under combustion like conditions. *Combust. Flame*, **102**, 418–426.

127 Atwood, A.I., Boggs, T.L., Curran, P.O., Parr, T.P., and Hanson Parr, D.M.

(1999) Burning rate of solid propellant ingredients. *J. Propul. Power*, **15**, 740–747.
128. Tang, K.C. and Brewster, M.Q. (2002) Modeling combustion of hydrazinium nitroformate. *Proc. Combust. Inst.*, **29**, 2897–2904.
129. Louwers, J., Gadiot, G.M.H.J.L., Landman, A.J., Peeters, T.W.J., van der Meer, Th.H., and Roekaerts, D. (1999) Combustion and flame structure of HNF sandwiches and propellants. AIAA Paper 1999–2359. 35th AIAA/ASME/SAE/ASEE Joint Propulsion Conference and Exhibit, Los Angeles, California, USA.
130. Parr, T.P. and Hanson-Parr, D.M. (1996) Solid propellant diffusion flame structure. 26th International Symposium on Combustion, Pittsburgh, 1981–1987.
131. de Klerk, W.P.C., van der Heijden, A.E.D.M., and Veltmans, W.H.M. (2001) Thermal analysis of the high-energetic material. *J. Thermal Anal. Calorim.*, **64**, 973–985.
132. Welland–Veltmans, W.H.M., Lillo, F., Del Cavaliere, C., van der Heijden, A.E.D.M., and Keizers, H.L.J. (2003) Current work on HNF based propellants in the perspective of future solid stages. AIAA publication no: IAC-03-S.2.02. 54th International Astronautical Congress of the International Astronautical Federation, the International Academy of Astronautics and the International Institute of Space Law, Bremen.
133. Ciucci, A., Froata, O., Welland, W., van der Heijden, A., Leeming, B., Bellerby, J., and Brotzu, A. (2004) Current state of the art of HNF composite propellants. Proceedings of the 2nd International Conference on Green Propellants for Space Propulsion (ESA SP-557), Sardinia, Italy.
134. Dendage, P.S., Sarwade, D.B., Asthana, S.N., and Singh, H. (2001) Hydrazinium nitroformate (HNF) and HNF based propellants: a review. *J. Energetic Mater.*, **19**, 41–78.
135. Sutton, G.P. and Biblarz, O. (2001) *Rocket Propulsion Elements*, 7th edn, John Wiley & Sons, Inc., New York.
136. Frisbee, R.H. (2003) Advanced space propulsion for the 21st century. *J. Propul. Power*, **19** (6), 1129–1154.
137. Gordon, S. and Mcbride, B.J. (1971) Computer program for calculation of complex chemical equilibrium compositions, rocket performance, incident and reflected shocks and Chapman-Jouguet detonations. *NASA SP*, **273**.
138. Gadiot, G.M.H.J.L., Mul, J.M., Meulenbrugge, J.J., Korting, P.A.O.G., Schnorhk, A.J. and Schoyer, H.F.R. (1993) New solid propellants based on energetic binders and HNF. *Acta Astronautica*, **29**, 771–779.
139. Publication of Munitions Safety Information Analysis Center (MSIAC). (2003) News Letter of 2nd Quarter Year. Internet Resource (http://www.msiac.nato.int).
140. Chan, M.L., Roy, E.M., and Turner, A. (1994) Energetic binder explosive. US Patent 5,316,600.
141. Radwan, M.A. (2006) Sensitivity and performance of energetic materials based on different types of energetic binders. Proceedings of 37th International Conference of the ICT, Karlsruhe, 56/1–56/12.
142. Meyer, R. (1981) *Explosives*, 2nd edn, Verlag Chemie, Weinheim.
143. Scilly, N.F. (1995) Measurement of explosive performance of high explosives. *J. Loss Prev. Process Ind.*, **8**, 265–273.
144. Cole, R.H. (1948) *Underwater Explosions*. Princeton University Press, Princeton, New Jersey, USA.
145. Graham, K.J. and Williams, E.M. (2002) Underwater explosive device. US Patent 6,354,220.
146. Bocksteiner, G. (February 1996) Evaluation of underwater explosive performance of PBXW-115 (Australia), Technical Report 0297, DSTO.
147. Keicher, T. Happ, A., Kretschmer, A., Sirringhaus, U., and Wild, R. (1999) Influence of aluminium/ammonium perchlorate on the performance of underwater explosives. *Propellants Explos. Pyrotech.*, **24**, 140–143.
148. Gaponik, P.N., Ivashkevich, O.A., Karavai, V.P., Lesnikovich, A.I.,

Chernavina, N.I., Sukhanov, G.T., and Gareev, G.A. (1994) Polymers and copolymers based on vinyl tetrazoles. Part 1(a): synthesis of poly (5-vinyl tetrazole) by polymer analogous conversion of polyacrylonitrile. *Die Angew. Makromol. Chem.*, **219**, 77–88.

149 Kizhnyayev, V.N., Kruglova, V.A., Ratovskii, G.V. Ye., Protasova, L., Vereshchagin, L.I., and Gareyev, G.A. (1986) Synthesis, study and chemical modification of vinyltetrazole polymers. *Polym. Sci. USSR*, **28**, 851–858.

150 Huang, M.-R., Li, X.-G., Li, S.-X., and Zhang, W. (2004) Resultful synthesis of polyvinyltetrazole from polyacrylonitrile. *React. Funct. Polym.*, **59**, 53–61.

151 Binder, W.H. and Kluger, C. (2006) Azide/alkyne – "click" reactions: applications in materials science and organic synthesis. *Curr. Org. Chem.*, **10**, 1791–1815.

152 Chafin, A. Irvin, D.J., Mason, M.H., and Mason, S.L. (2008) Synthesis of multifunctional hydroxyethyl tetrazoles. *Tetrahedron Lett.*, **49**, 3823–3826.

153 Demko, Z.P. and Sharpless, K.B. (2001) Preparation of 5-substituted 1H-tetrazoles from nitriles in water. *J. Org. Chem.*, **66**, 7945–7950.

154 Miller, C.G. and Williams, G.K. (2006) Water based synthesis of poly (tetrazoles) cross-reference to related applications. International Publication Number WO 2008/059318 A2.

155 Roshchupkin, V.P., Nedel'ko, V.V., Larikova, T.S., Kurmaz, S.V., Afanas'yev, N.A., Fronchek, E.V., and Korolev, G.V. (1989) Thermally induced transformations in the homologous series of polymers of 5-vinyltetrazole and its 5-alkyl derivatives. *Polym. Sci. USSR*, **31**, 1900–1908.

156 Mishra, I.B. and Vande Kieft, L.J. (1989) Insensitive binder for propellants and explosives. US Patent 4,875,949.

157 Mishra, I.B. and Vande Kieft, L.J. (March 1989) Polyethylene glycol – poly (2-methyl-5-vinyl tetrazole) polymer blend (a desensitizing binder for propellants and explosives). Technical Report No. BRL-TR-2985, Ballistics Research Laboratory, Aberdeen Proving Ground, Maryland, USA.

158 Steinhauser, G. and Klapotke, T.M. (2008) "Green" pyrotechnics: a chemist's challenge. *Angew. Chem. Int. Ed.*, **47**, 3330–3347.

159 Singh, R.P., Gao, H., Meshri, D.T., and Shreeve, J.M. (2007) Nitrogen-rich heterocycles, in *High Energy Density Materials, Structure and Bonding*, **Vol. 125** (ed. T.M. Klapotke), Springer-Verlag, pp. 35–83.

160 (a) Yoshizawa, M., Ogihara, W., and Ohno, H. (2002) Novel polymer electrolytes prepared by copolymerization of ionic liquid monomers. *Polym. Adv. Technol.*, **13**, 589–594; (b) Tang, J., Sun, W. Tang, H., Radosz, M., and Shen, Y. (2005) Enhanced CO_2 absorption of poly(ionic liquid)s. *Macromolecules*, **38**, 2039–2037.

161 Edward, T.R. and Sidney, S.G. (1968) Triaminoguanidinium salts of 5-vinyltetrazole polymers and a method for their preparation. US Patent 3,397,186.

162 Xue, H., Gao, H., and Shreeve, J.M. (2008) Energetic polymer salts from 1-vinyl-1,2,4-triazole derivatives. *J. Polym. Sci. A: Polym. Chem.*, **46**, 2414–2421.

163 (a) Kizhnyaev, N.V., Vereschchagin, L.I., Verkhozina, O.N., Pokatilov, F.A., Tsypina, N.A., Petrova, T.L., Sukhanov, G.T., Gareev, G.A., and Smirnov, A.I. (2003) Triazole and tetrazole containing energetic compounds. Proceedings of 34th International Annual Conference of ICT, Karlsruhe, 75/1–75/11; (b) Kizhnyaev, N.V., Petrova, T.L., and Smirnov, A.I. (2001) Rheological properties and gel formation of aqueous salt containing solutions of sodium poly (5-vinyltetrazolate) in the presence of Cr^{+3} ions. *Polym. Sci. Ser. A*, **43**, 566–571.

164 Filter, H.E., Bidlack, H.D., and Stevens, D.L. (1975) Solid propellant composition with aziridine cured epichlorohydrin polymer binder. US Patent 3,890,172.

165 Jain, S.R., Sridhara, K., and Thangamatesvaran, P.M. (1993) N–N bonded polymers. *Prog. Polym. Sci.*, **18**, 997–1039.

166 Amanulla, S. and Jain, S.R. (1997) Synthesis and characterization of epoxy binders based on N,N′-aliphatic dicarboxyl bis(hydrazones). *J. Polym. Sci. A Polym. Chem.*, **35**, 2835–2842.

167 Rajendran, G. and Jain, S.R. (1984) Thermal analysis of monothiocarbonohydrazones. *Thermochim. Acta*, **82**, 311–323.

168 Oommen, C., Amanulla, S., and Jain, S.R. (2000) Characterization of diglycidylamine epoxy resins based on bis-hydrazones. *Eur. Polym. J.*, **36**, 779–782.

169 Provatas, A. (2000) Energetic polymers and plasticizers for explosive formulations – a review of recent advances, Technical Report 0966, DSTO.

3
Nitropolymers as Energetic Binders

3.1
Introduction

Nitropolymers are a unique class of energetic polymers containing nitrato ($-ONO_2$) pendant groups attached to the backbone of the polymer chain. These polymers differ from the high nitrogen content polymers described in the previous chapter through the mechanism of energy release. In high nitrogen content polymers, energy is released by the highly exothermic scission of the $-N-N-$ bond in the backbone or in the pendant groups of the polymer. In the case of nitropolymers, energy is released by the oxidation of the polymer backbone chain, promoted by the scission of the $-O-NO_2$ bond in the nitrato-pendant groups. Therefore, the same nitropolymer molecule contains both the fuel and oxidizer components to function as an energetic material.

This feature of nitropolymers was exploited by Paul Vieille as early as 1886 through the use of the nitropolymer nitrocellulose (NC) as a single base propellant for use in gas generators and ejectors. The performance parameters of single-base propellants were further improved in 1887 (Ballistite, Alfred Nobel) by the additional energetic ingredient nitroglycerin (NG) to form extruded double base propellants, which are more adapted to rockets. Over the years, with the increasing demand for higher performance and improved mechanical strength, the NC-based propellant family has evolved from extruded double base to elastomeric modified cast double-base propellants.

With the advent of the heterogeneous branch of the propellant family in 1950, newer energetic nitropolymers, poly(glycidyl nitrate) (PGN), poly(3-nitratomethyl-methyl oxetane) (Poly(NIMMO)), and nitrated hydroxy terminated polybutadiene (NHTPB), were developed. These are hydroxyl terminated prepolymers to function as energetic binders for heterogeneous propellants.

This chapter details the preparation, properties, and applications of nitropolymers as energetic binders for propellant and explosive formulations.

Energetic Polymers: Binders and Plasticizers for Enhancing Performance, First Edition.
How Ghee Ang and Sreekumar Pisharath.
© 2012 WILEY-VCH Verlag GmbH & Co. KGaA, Weinheim. Published 2012 by WILEY-VCH Verlag GmbH & Co. KGaA

3.2
Preparation of Nitropolymers

3.2.1
Nitrocellulose

Nitrocellulose (NC) is one of earliest of the known energetic materials [1]. A brief historical perspective about the evolution of NC as an energetic material is presented in the introductory chapter of this book. Generally, NC is prepared by the nitration of cellulose, a linear polymer of β-1,4-linked D-glucopyranose units (Scheme 3.1) [1].

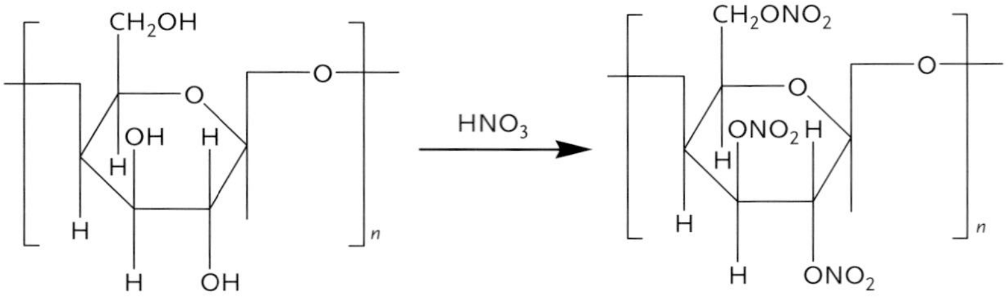

Scheme 3.1 Preparation of nitrocellulose.

The natural sources of cellulose are cotton and wood. Molecular heterogeneity of the prepared NC depends on the source material. Molecular weight measurements demonstrate that the NCs prepared from wood are distinctly more heterogeneous than those prepared from cotton [2]. The greater heterogeneity of wood nitrocellulose is ascribed to the high content of low molecular weight material in wood. A sulfonitric mixture consisting of 63.5% H_2SO_4, 22% HNO_3, and 14.5% H_2O is used for nitration.

The properties and application areas of NC vary with the percentage of nitration. At a low percentage of nitration, NC forms a tough thermoplastic film that is useful as a lacquer and has importance in the photographic industry. At a high percentage of nitration (ranging from 11.5 to 14%), NC is used as a major ingredient in solid propellants. The most frequently employed NC has a nitrogen content of 12.6%. Thermochemistry, and consequently the ballistic behavior of NC, is a function of the percentage of nitrogen in the polymer. Both heat of formation and calorific value of NC increases linearly with the nitrogen content of the polymer [3–5]. For example, the heat of formation of NC increases from −2.6 to −2.3 kJ/g when the nitrogen content of the polymer increases from 12.1 to 13.24% [3]. As cellulose is a natural polymer, characteristics of the prepared NC differ with the geographical origin and seasonal variations of the starting cellulose. Hence, implementation of a rigid product quality control is mandatory during the manufacture of NC, in order to obtain products with predictable properties. Over the years, reliable predictive models have evolved, which are independent of the initial molecular weight distribution of the cellulose used, to accurately predict the nitrogen content of polymer with respect to reaction variables [6].

3.2.2
Poly(Glycidyl Nitrate) (PGN)

Hydroxy-terminated energetic prepolymers were beginning to be explored as starting materials for propellants/explosives in 1947 with the advent of the composite propellant family. Within this category, polymers of monomer glycidyl nitrate (GN or GLYN) were the first to be investigated by Thelan and coworkers as early as 1950 at the Naval Warfare Center, USA [7]. However, the recognition of Poly(GN), as an energetic binder had to wait for another 40 years, because of the hazardous chemical routes involved in the synthesis of GN.

Traditionally, GN was prepared by the nitration of glycidol using a potentially dangerous nitrating mixture consisting of acetic anhydride and 95% nitric acid. A cumbersome purification process is required to recover GN from the reaction mixture [7]. In the early 1990s, the Defence Research Agency (DRA), in the United Kingdom, exploited the utility of dinitrogen pentoxide (N_2O_5) as a nitrating agent, and established a safe process to prepare GN monomer solutions in high yield and purity (Scheme 3.3) from glycidol [8].

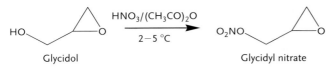

Scheme 3.2 Preparation of GN monomer using a nitrating mixture.

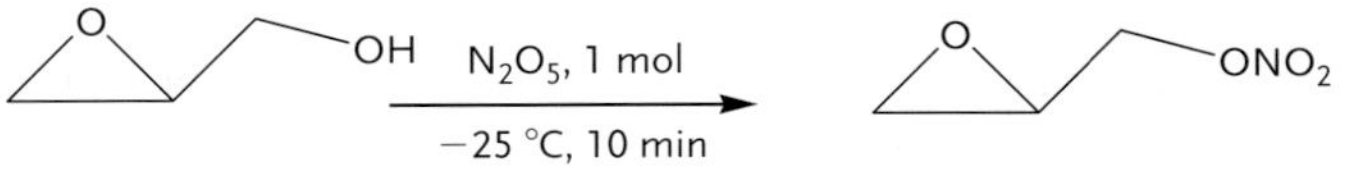

Scheme 3.3 Preparation of GN monomer using dinitrogen pentoxide.

GN prepared using N_2O_5 does not require any further purification before proceeding to the polymerization step towards the preparation of PGN. The reaction conditions should be carefully controlled to avoid the production of the highly explosive nitroglycerin during the reaction [9]. Glycidol should be purified by distillation before proceeding to the nitration, to ensure complete conversion of glycidol into GN.

A new route for the synthesis of GN has been developed by ATK Thiokol [10], in which distillation or vaporization techniques are not required prior to the polymerization of GN (Scheme 3.4).

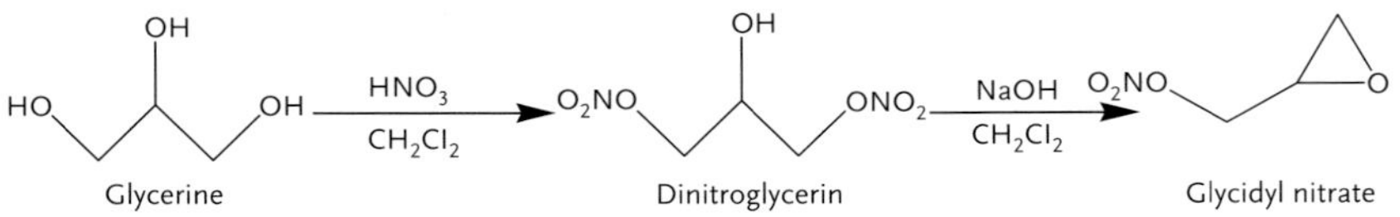

Scheme 3.4 Preparation of GN monomer from glycerin.

The nitroglycerin and dinitroglycerin formed as by-products during the nitration step can be carried to the polymerization stage along with GN and retained with PGN as energetic ingredients.

The polymerization of GN into PGN is achieved by cationic ring opening polymerization, in the presence of an acid catalyst and polyfunctional alcohol (polyol) initiator (Scheme 3.5).

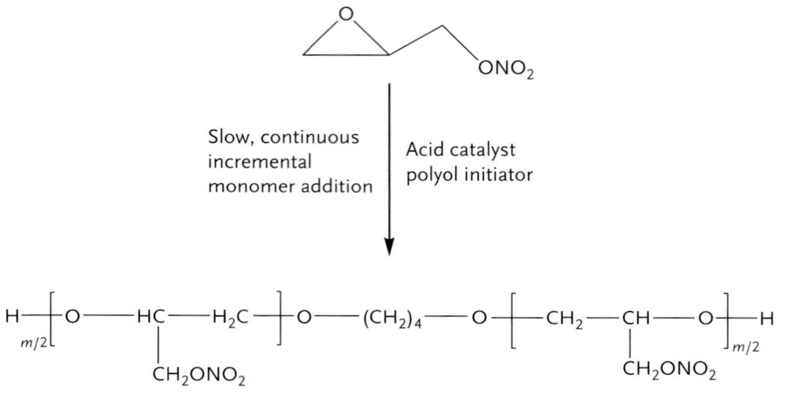

Scheme 3.5 Reaction scheme for the polymerization of GN.

For a given monomer concentration, polymers with differing molecular weights and hydroxyl values could be obtained by varying the relative amount of the acid catalyst to initiator. The conditions are controlled in such a way that the reaction proceeds by the preferred activated monomer mechanism (AMM) over the conventional active chain end (ACE) mechanisms. Briefly, AMM and ACE are the two important mechanisms involved in the cationic activated monomer polymerization of heterocyclic monomers such as GN. AMM involves successive additions of protonated monomer molecules to terminal hydroxyl groups of the growing polymer chains, whereas ACE involves reaction of monomer molecules with tertiary oxonium active species located at the chain ends [11]. The important feature of AMM polymerization is that it is quasi-living, which serves to provide hydroxyl terminated polymers with low polydispersity. Furthermore, as the growing polymer chain does not carry any charge, the possibility of back-biting reactions resulting in the formation of low molecular weight cyclic products is fully eliminated; that is, the polymer becomes free of any oligomer contamination [12, 13].

Therefore, AMM is the preferred mechanism over ACE in energetic polymer syntheses for binder applications.

Acid catalysts employed for the polymerizations of GN to PGN are protic and Lewis acids. They include the Lewis acids $BF_3 \cdot Et_2O$, $BF_3 \cdot THF$, $BF_3 \cdot HBF_4$, BF_3 (gas), and PF_5, and the protic agent HBF_4 and triethoxonium salts (triethoxonium hexafluorophosphate, triethoxonium hexafluoroantimonate, and triethoxonium tetrafluoroborate) [10, 14, 15]. The polyol initiators are usually diols: 1,4-butanediol and ethylene glycol. In order to obtain polymers with higher functionality, triols (trimethylol propane) and tetrols (2,2′-dihydroxymethyl-1,3-propanediol) are used as initiators [10(b)]. It was discovered that controlled reaction occurs if the molar ratio of the acid catalyst to that of the diol is kept in the range of 0.4 : 1 to 0.8 : 1. By using a substantially lower amount of the acid catalyst, incorporation of a larger proportion of the polyol molecules within the polymer is achieved [14].

Earlier investigations on PGN and PGN propellants at the Jet Propulsion Laboratory (JPL) indicated that PGN made with the widely used acid catalyst $BF_3 \cdot Et_2O$ was low in both functionality and molecular weight, and therefore polyurethane propellants made from this PGN have inferior mechanical properties [14, 16]. Later, experimental trials at DRA pointed out that tetrafluoroboric acid and 1,4-butanediol offers a good catalyst combination to obtain low disperse, di- or tri-hydroxyl end functional PGNs in a range of molecular weights (500–3500) [17]. A slow rate of monomer addition (over a period of 16–40 h) is mandatory to drive the polymerization through the AMM route, which serves to produce polymers free of low molecular weight cyclic contaminants [17].

PGN produced by the above mentioned processes produces polymer with atactic stereochemistry. As a result, the synthesized PGN is a highly viscous amorphous liquid that must be cured with multifunctional polyisocyanates to provide binders for solid propellants. In view of the inherent disadvantages of cross-linked elastomeric polymers as binder materials, a process was developed to produce crystalline isotactic PGN that could be used as a hard block for PGN based thermoplastic elastomer materials [18] (Scheme 3.6). Formulations based on thermoplastic elastomer materials provide the advantage of recycling. The topic of energetic thermoplastic elastomers will be discussed in more detail in Chapter 4.

In this process, isotactic PGN was synthesized by polymerizing chiral (R) glycidyl nitrate or its enantiomer (S) glycidyl nitrate. Chiral (R or S) glycidyl nitrate is prepared by either the direct nitration of R- or S-glycidol or by the treatment of chiral glycidyl tosylate with nitric acid and sodium hydroxide. The chiral glycidyl nitrate is polymerized to obtain an isotactic PGN with a melting-point of 47.2 °C.

3.2.2.1 Curing of PGN

Similar to the azido polymers, the hydroxyl terminated PGN is converted into an energetic polyurethane binder by the reaction of the terminal hydroxyl groups with aromatic or aliphatic polyisocyanates. However, the cured PGNs undergo self-decomposition or de-curing (causing chain scissions) at room temperature

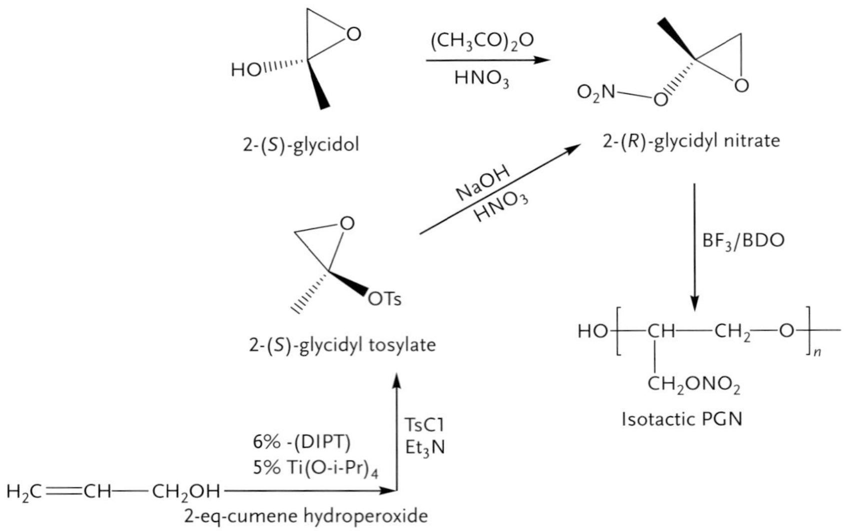

Scheme 3.6 Process for producing isotactic PGN.

due to the proximity of the terminal hydroxyl groups of the polymer to the nitrate ester groups (Scheme 3.7) [19].

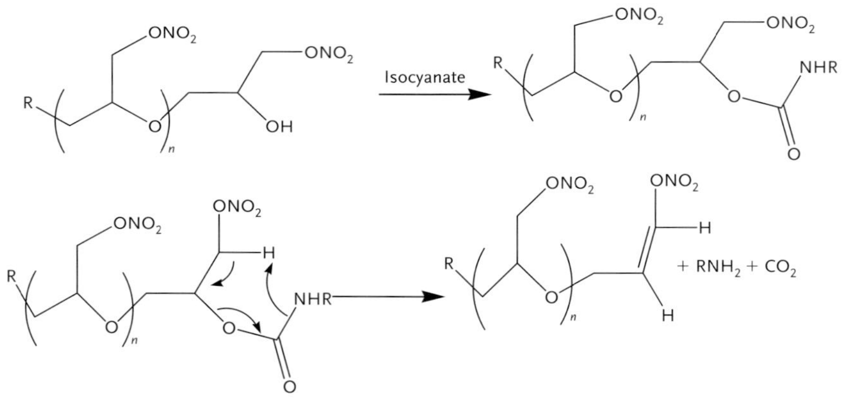

Scheme 3.7 Scheme of de-curing of PGN binder.

Several approaches have been suggested for solving the de-curing problem of PGN. Sanderson et al. [19] prepared a polyfunctional PGN having functionality greater than three using a multifunctional alcohol with a hydroxyl functionality of at least two and cured it with an aromatic diisocyanate, which improved the ageing properties of the binder.

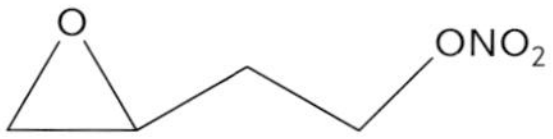

Figure 3.1 Structure of 2-nitratoethyl oxirane.

Kim and coworkers modified the structure of GN to 2-nitratoethyl oxirane (Figure 3.1) by inserting an extra methylene group [20], which resulted in a new energetic prepolymer.

Synthesis of 2-nitratoethyl oxirane and its polymerization to poly(2-nitratoethyl oxirane) is presented in Scheme 3.8.

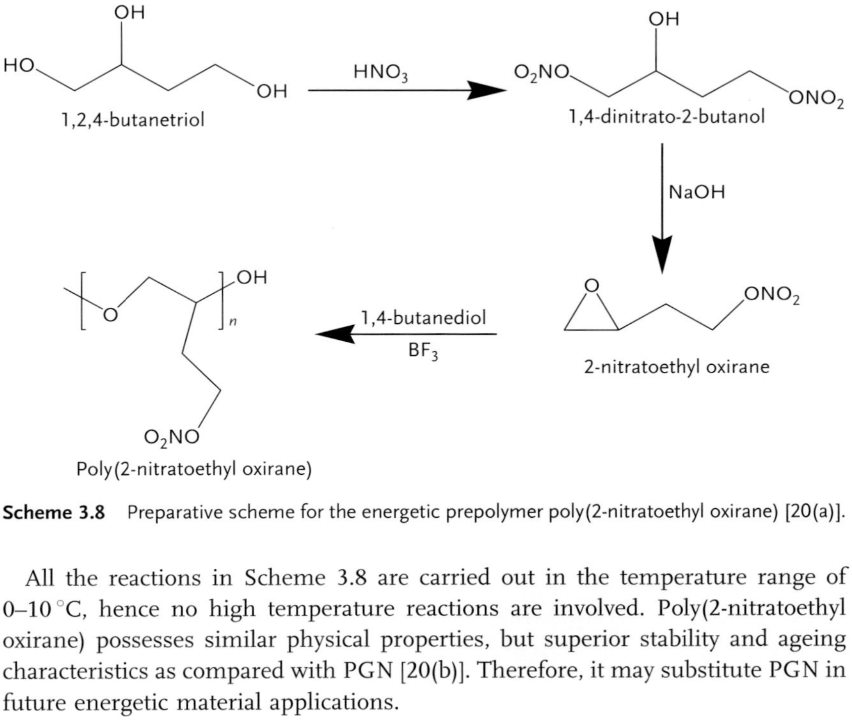

Scheme 3.8 Preparative scheme for the energetic prepolymer poly(2-nitratoethyl oxirane) [20(a)].

All the reactions in Scheme 3.8 are carried out in the temperature range of 0–10 °C, hence no high temperature reactions are involved. Poly(2-nitratoethyl oxirane) possesses similar physical properties, but superior stability and ageing characteristics as compared with PGN [20(b)]. Therefore, it may substitute PGN in future energetic material applications.

Paul et al. at ICI [20(c)] proposed a two-step method of stabilization of PGN by epoxidation of the terminal hydroxyl group and the adjacent nitrate ester, followed by opening of the epoxide ring using H_2SO_4, thus providing a terminal hydroxyl group in the place of the original nitrate ester. Later Paraskos et al. [20(d)] improved the method and accomplished the end-group modification in a single step. This approach was unable to fully solve the de-curing problem. Moreover, the energetic performance of the binder drops due to the removal of at least 10% of the nitrate ester groups in this technique [19]. Research work at ICI showed that the most effective way of removing the instability of PGN is to treat the end-modified PGN with potassium carbonate [20(e)].

3.2.3
Poly(Nitratomethyl-methyl Oxetane) (Poly(NIMMO))

Poly(NIMMO) is a versatile energetic oxetane polymer from the nitropolymer family. It is a hydroxyl terminated homopolymer of the energetic 3-nitratomethyl-3-methyl oxetane (NIMMO) monomer and can be cured by isocyanates to give energetic binders.

As with PGN, the main challenge in the development of Poly(NIMMO) as an energetic binder was the identification of a safe and economic synthetic route for the preparation of the NIMMO monomer. Several synthetic schemes have been reported in the literature for the synthesis of NIMMO, Scheme 3.9 [21–23].

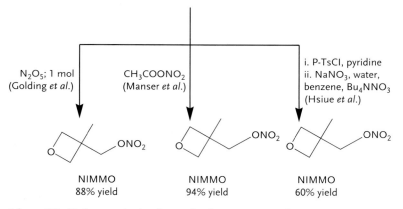

Scheme 3.9 Various synthetic schemes for the preparation of NIMMO.

The common starting oxetane derivative for the preparation of NIMMO is 3-methyl-3-hydroxymethyl oxetane (HMMO), which is synthesized by reaction of

2-methyl-2-hydroxymethyl-propanediol with diethyl carbonate followed by cyclization using alcoholic potassium hydroxide [24].

Even though the method according to Manser [22] resulted in excellent yields of NIMMO, it required the preparation of acetyl nitrate by reacting acetic anhydride with concentrated nitric acid. Acetyl nitrate is a very hazardous and unstable explosive that can detonate spontaneously, which makes the industrial level scale-up of this process difficult and unsafe. The method reported by Hsiue and coworkers [23] uses toxic chemicals, such as pyridine, and involves multiple steps, resulting in the process becoming expensive and commercially unviable. The simple and efficient method of preparing NIMMO by the direct liquid phase nitration of HMMO using dinitrogen pentoxide [21] produces high-purity monomer, which does not require any further purification procedures. This process has been successfully scaled-up by DRA and the technology transferred to ICI for commercial production of Poly(NIMMO) [25].

NIMMO is polymerized to the hydroxyl terminated energetic polyether by cationic ring opening polymerization using acid catalysts and polyfunctional initiators (Scheme 3.10).

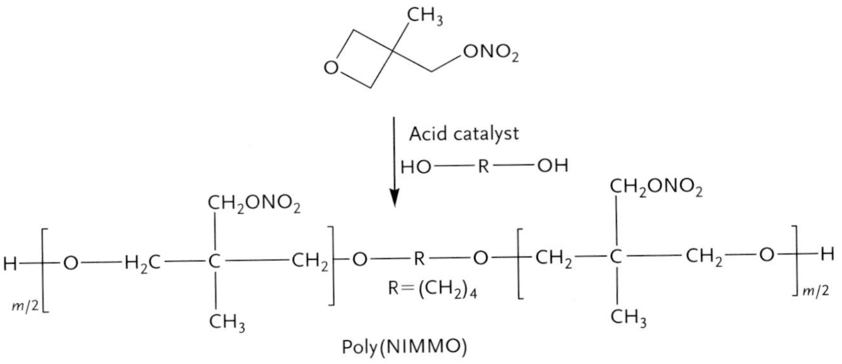

Scheme 3.10 Preparation of Poly(NIMMO).

The acid catalysts preferably used for the polymerization of NIMMO include $BF_3 \cdot Et_2O$, HBF_4, $HSbF_6$, and $BF_3 \cdot THF$ [26]. Usually methylene chloride is employed as the solvent for polymerization. Also, significant amounts of water or brine are used in the termination step of the polymerization, which becomes contaminated with the energetic polymer and generates hazardous waste. Farncomb and coworkers [27] successfully used liquid carbon dioxide as a solvent to replace the toxic methylene chloride for the polymerization of NIMMO. The liquid carbon dioxide medium acts as a heat sink for the reaction and also as a separating medium (Poly(NIMMO) is virtually insoluble in liquid carbon dioxide) for the nitrated chemical compositions. Furthermore, the carbon dioxide process uses only water, but in much smaller amounts, thus minimizing the hazardous waste generation.

3.2.4
Nitrated HTPB (NHTPB)

Nitrated derivatives have been synthesized from HTPB by a combination of epoxidation and nitration reactions (Scheme 3.11) [9, 28].

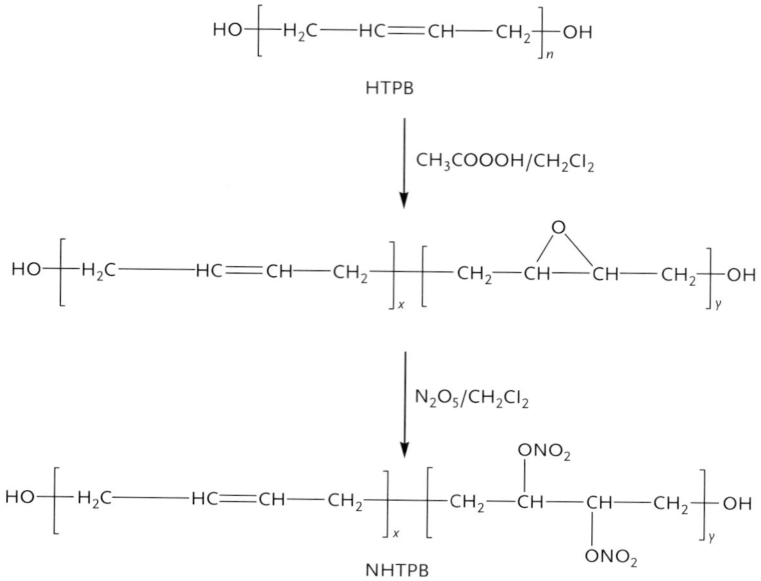

Scheme 3.11 Preparation of nitrated HTPB [9].

The properties of NHTPB depend on the percentage of the double bonds converted into the nitrate ester groups. The level of nitration affects the thermal stability and mechanical properties of the polymer. To obtain a good compromise between various properties, a level of 10–15% nitration is suggested [29]. NHTPB has a low viscosity for a processing environment and can be cured with aliphatic and aromatic isocyanates. Unlike the HTPB, NHTPB offers the advantage of miscibility with energetic plasticizers, which could be utilized for improving the energy of PBX and propellants [28].

3.2.5
Nitrated Cyclodextrin Polymers (Poly(CDN))

Cyclodextrins (CD) are cyclic molecules consisting of D-glucose units with hydrophilic hydroxyl groups lying along the rim of the molecule while the interior cavity of the molecule is non-polar and lipophilic. Each D-glucose in a CD has three free –OH groups capable of being nitrated to a nitrate ester group, to form energetic nitrated cyclodextrin polymers (CDN) (Figure 3.2) [30].

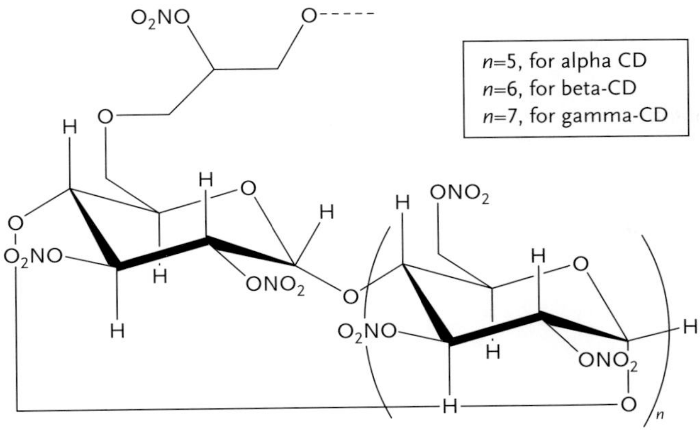

Figure 3.2 Nitrated cyclodextrin polymer cross-linked with 1-chloro-2,3-epoxy propane.

In order to function as efficient binders, CDNs should have high molecular weights. Before nitration, the CDs are cross-linked to high molecular weight polymers by adopting one of the following linking methods [31]:

1. hydroxypropyl linkages using 1-chloro-2,3-epoxy propane;
2. urethane linkages using 4,4′-methylene-bis(phenyl isocyanate);
3. attached to linear polymers such as poly(allyl amine) as pendant groups.

After cross-linking, the CDs are nitrated by the standard nitration technique of using nitric acid or by employing N_2O_5 in liquid carbon dioxide to obtain Poly (CDN). The route using N_2O_5 is suitable for CDs with polyurethane (PU) linkages, as PU linkages are susceptible to hydrolysis by nitric acid.

The lipophilic cavity of the CDN encapsulates a wide variety of energetic materials including nitrate esters and nitramines to form molecular explosive complexes [31(d)]. The complex is formed by dipole–dipole interaction. As the energetic material is contained in the lipophilic core of the inclusion complex, the CDN acts as a protective shield for the energetic material and resists its accidental detonation due to impact, shock, friction, or electric discharge. Furthermore, CDN as such adds in additional energy, which improves the overall performance of the complex. Hence, CDN complexes offer potential as high performance insensitive munition energetic candidates.

3.3
Thermal Decomposition Behavior of Nitropolymers

3.3.1
Nitrocellulose

The thermal decomposition behavior of nitrocellulose polymers has been extensively studied over the years by using a variety of techniques [32–35]. Under slow

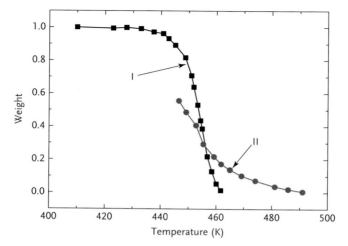

Figure 3.3 TGA curve of 5 mg NC in air at heating rate of 0.5 K/min. Values for the plot taken with permission of Elsevier from Ref. [34]. The first part of the curve (I) obeys simple first order kinetic rate law and the second part (II) autocatalytic rate law.

heating rates ($dT/dt < 5\,°C/min$, the kinetics are controlled by first-order autocatalysis and simple first order separated by a discontinuity (Figure 3.3). The first-order autocatalysis law is defined by:

$$\frac{d\alpha}{dt} = k_1(1-\alpha) + k_2\alpha(1-\alpha) \qquad (3.1)$$

where

K_1 and K_2 are rate constants
α is the fractional conversion.

Among the rate constants, the slower one is assigned to the first-order homolytic cleavage of the O–NO$_2$ bond of NC and the faster to the autocatalytic stage.

The kinetic data reported in the literature for the thermal decomposition of NC at slow heating rates, obtained by various methods, are compiled in Table 3.1. Because of the differences in the weight of samples used and the nature of the measurements, there is significant scatter in the activation energy values obtained for the first-order stage ($E_a = 130–178$ kJ/mol) and for the autocatalytic stages ($E_a = 168–200$ kJ/mol). The activation energies of thermal decomposition for both the stages of NC are practically invariant with respect to the percentage of nitrogen (%N).

The Arrhenius plots for the three NC samples with different %N (Figure 3.4) show first-order behavior at lower temperature, with autocatalysis becoming important at higher temperatures. The rate laws determining thermal decomposition remain unchanged with the %N.

Even though there are no obvious differences in the kinetic parameters among the NC samples with different %N, the extent of autocatalysis involved in the

Table 3.1 Kinetic parameters for the thermal decomposition of NC determined using various techniques at slow heating rates.

Method		E1 (kJ/mol) (first order)	Ln A1 (1/s)	E2 (kJ/mol) (autocatalytic)	Ln A2 (1/s)	Reference
Iso-DTA		130	27.6	200.13	43.8	[36]
TGA		178.36	38.2	175.43	40.3	[33(b)]
IR		174.17	38.23	166.63	38.23	[34]
Raman		158.68	35.46	175.43	39.15	[34]
TGA of NC with different %N	%N	E1 (kJ/mol)	Ln A1 (1/s)	E2 (kJ/mol)	Ln A2 (1/s)	[35(a)]
	13.4	169.98	34.8	171.66	36.85	
	11.7	170.82	35.0	172.08	36.85	
	9.45	174.59	35.5	172.5	36.62	

decomposition process is determined by the ratio of primary to secondary alkyl nitrate groups in the molecule. This ratio is highest in NC with 9.45%N and lowest in the sample with 13.4%N. Autocatalysis occurs to a lesser extent in NC with 9.45%N as compared with that having 13.4%N, suggesting that the autocatalysis is favored by the presence of the secondary nitrate ester groups [35(b)]. The molecular basis of the autocatalytic stage of NC remains unknown. It is presumed that the main reason for the autocatalytic reaction is the development of oxidative and hydrolytic interaction of NC with products of its decomposition: HNO_3, NO_2, and H_2O [35(c), 36(b)]. Because of this autocatalytic reaction, stabilizers are a must for propellant and explosive formulations using NC polymer binders.

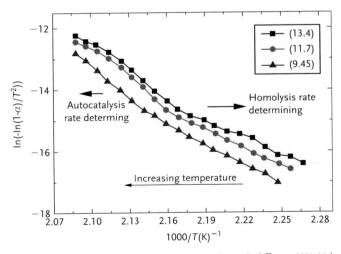

Figure 3.4 Arrhenius plots for three NC samples with different %N. Values for the plot taken with permission of Elsevier from Ref. [35(a)].

Table 3.2 Kinetic parameters for thermal decomposition of NC at high heating rates.

Method	%N	E (kJ/mol)	Ln A (1/s)	Reference
SMATCH–FTIR	9.45	141.51	38.46	[35(b)]
	11.7	141.10	38.92	
	13.4	136.49	37.77	
T-jump	12.4	126.86	27.0	[35(c)]
	12.6	121.84	26.4	
	13.2	113.88	24.7	
	13.4	81.64	16.6	

An IR spectrum of the product isolated at the discontinuity in the TGA (thermogravimetric analysis) curve (Figure 3.3) identifies the product as a fragmented type of nitrated oxycellulose [37]. The basic skeleton of NC is unchanged during the autocatalytic decomposition [33(a), 37]. After the autocatalytic run out, the reaction proceeds by simple kinetics, with activation energy of the order of magnitude of the cleavage of the O–NO$_2$ bond [34]. This carbonization reaction degrades the intermediate to the charcoal type material. The products of the autocatalytic and the ensuing first-order steps are also observed in combustion reactions.

The kinetic parameters obtained at high heating rates for the thermal decomposition of NC are summarized in Table 3.2. At faster heating rates (>100 °C/s), the Arrhenius rate law ($d\alpha/dt = A(1-\alpha)^n \exp(-E_a/RT)$) linearizes with $n = 2$ [35(b)]. This does not mean that the reaction order is necessarily second order in the usual sense of chemical kinetics, and $n = 2$ is only an empirical parameter to linearize the power rate law. The important point to note is that the reaction rate ($d\alpha/dt$) and the function describing the degree of conversion ($f(\alpha)$) follow different relationships at slow and fast heating rates [35(c)].

The combustion reaction of NC starts with the cleavage of one of the O–NO$_2$ nitrate ester bonds as illustrated in Scheme 3.12 [33(a), 37].

Two possible pathways of the scission of the O–NO$_2$ bonds are: (i) from the nitrate group attached to C3 of the cellulose, and (ii) from the nitrate group attached to C6 of the cellulose. The first route (see route A of Scheme 3.12) leads to the opening of the glucose ring and the second one (route B) leads to the formation of aldehyde.

The aldehyde molecules then react with NO$_2$ to produce NO, which oxidizes the organic molecules exothermically producing combustible CO, CO$_2$, N$_2$, CH$_4$, HCHO, and H$_2$O. Combustion of NC-based propellants will be discussed in more detail in a later section.

3.3.2
Poly(Nitratomethyl-Methyl Oxetane) (Poly(NIMMO))

Thermal decomposition of the energetic nitropolymer Poly(NIMMO) at slow heating rates have been investigated by differential scanning calorimetry, TGA,

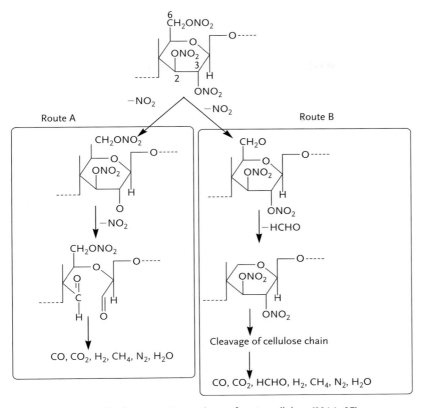

Scheme 3.12 Possible decomposition pathways for nitrocellulose [33(a), 37].

adiabatic calorimetry, and heat flow calorimetry [35(a), 38]. In all instances the polymer decomposition follows a simple first-order rate law. The kinetic parameters for slow heat rate thermal decomposition of Poly(NIMMO) obtained by various techniques are summarized in Table 3.3.

The autocatalytic behavior of the thermal decomposition of NC shifts to a simple first-order decomposition when it comes to Poly(NIMMO). It is deduced that three

Table 3.3 Kinetic parameters for the thermal decomposition of Poly(NIMMO) at low heat rates.

Method	E (kJ/mol)	Ln A (1/s)	Reference
Isothermal differential calorimetry	163.67	35.66	[38]
Heat flow calorimetry (HFC) (under nitrogen)	156.75 ± 4.4	NA	[38]
Adiabatic calorimetry (ARC)	180.2 ± 2.93	40.25	[38]
Thermogravimetric analysis (TGA)	185.06	41.45	[35(a)]

factors are responsible for the difference in kinetic behavior between NC and Poly (NIMMO) [35(a)]. They are the differences in phase, oxygen balance, and the number of primary and secondary nitrate groups. Firstly, with regard to the phase, NC is an amorphous solid at room temperature whereas Poly(NIMMO) is a viscous liquid. The nature of the phase determines the extent of the build-up of the concentration of NO_2, which is responsible for the autocatalytic behavior. Because Poly(NIMMO) is a viscous liquid, the NO_2 build up does not happen to the level required for autocatalysis, as compared with that in a more rigid solid NC matrix.

The second projected explanation of the difference in kinetic behavior is the lower oxygen balance (OB) of Poly(NIMMO) as compared with that for NC. OB determines the amount of NO_2 available relative to other decomposition products. The higher the OB, the higher is the amount of NO_2 available. The acceleratory effect on the decomposition of the sample requires a threshold amount of NO_2 being reached. Because of its lower oxygen balance, Poly(NIMMO) requires a higher threshold amount of NO_2 to initiate the autocatalytic reaction at higher temperatures compared with NC.

Thirdly, kinetic behavior is determined by the nature of nitrate ester groups in the backbone of the sample. As mentioned in Section 3.3.1, autocatalysis is favored by the presence of secondary alkyl nitrate groups. Poly(NIMMO) posses only primary alkyl nitrate groups in its backbone. Hence, it follows simple first-order kinetics for thermal decomposition.

Thermal decomposition of Poly(NIMMO) at fast heating rates (exceeding 100 °C/s) has been studied by SMATCH–FTIR spectroscopy [35(b)]. The decomposition behavior of Poly(NIMMO) obeys second-order kinetics with $E_a = 140.26 \pm 1.5$ kJ/mol and $\ln A = 37.77 \pm 0.5$ (1/s), which is similar to that obtained for NC. The similarities in the kinetic parameters for thermal decomposition of nitrate ester polymers suggest that similar overall reactions and transport properties dominate the decomposition process at high heating rates.

Mechanistic studies on the thermal decomposition mechanism of Poly (NIMMO) are few. Kemp et al. [39] exhaustively studied the thermal decomposition of Poly(NIMMO) using a variety of spectroscopic methods, identified the products, and outlined the mechanism in terms of those identified for other types of polyesters. Size-exclusion chromatography (SEC) shows that both chain scission and cross-linking reactions take place during the pyrolysis, and the dominance of each process in pyrolysis depends on the availability of oxygen. Presumably in the presence of excess of oxygen, the rate of oxidation and thus of chain scission is far greater than that of cross-linking. The importance of the latter increases as the oxygen supply becomes exhausted. A significant observation is the appearance and gradual increase in intensity of absorptions around 1729 and 1550 cm^{-1} in the solution IR spectrum of Poly(NIMMO) due to thermal degradation. The band at 1729 cm^{-1} is ascribed to the formation of formate ester due to the C–C bond scissions, preceded by the oxidation of main chain methylenes (Scheme 3.13).

The absorption at 1550 cm^{-1} is attributed to the asymmetric stretch of a nitro group attached to the tertiary carbon atom (a tertiary nitroalkane), which is formed by the recombination of NO_2 with the carbon radical resulting from the elimination

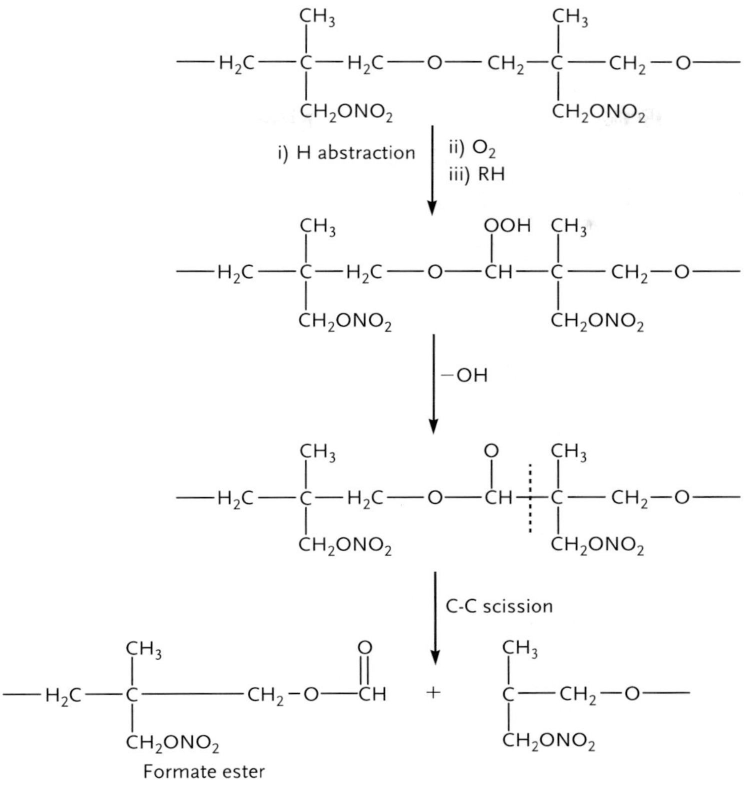

Scheme 3.13 Formation of formate ester from decomposition of Poly(NIMMO) [39].

of NO_2 and CH_2O from the Poly(NIMMO) side chains (Scheme 3.14). It should be noted that the formation of tertiary nitroalkane is not a result of any chain scissions.

The mechanisms proposed in Scheme 3.14 have been substantiated by the presence of significant amounts of formaldehyde molecules in the pyrolysis gases of Poly(NIMMO) and the gradually increasing concentration of the nitroxide radical detected by the strong enhancement of the electron spin resonance signal. The nitroxide radical is formed by the spin-trapping of alkyl radicals by the tertiary nitroalkane as shown in Eq. (3.2) [39].

$$R\bullet + NO_2 \rightarrow R - NO_2 \xrightarrow{R'\bullet} R - N(-O^\bullet) - OR' \qquad (3.2)$$

Comparing thermal decomposition mechanisms of Poly(NIMMO) and NC, it could be observed that the chemical structure of Poly(NIMMO) is more prone to chain scissions during decomposition (Scheme 3.13), which causes deterioration of the mechanical properties of the Poly(NIMMO) binder. On the other hand, the NC molecule retains the weight-bearing skeleton even during the autocatalytic

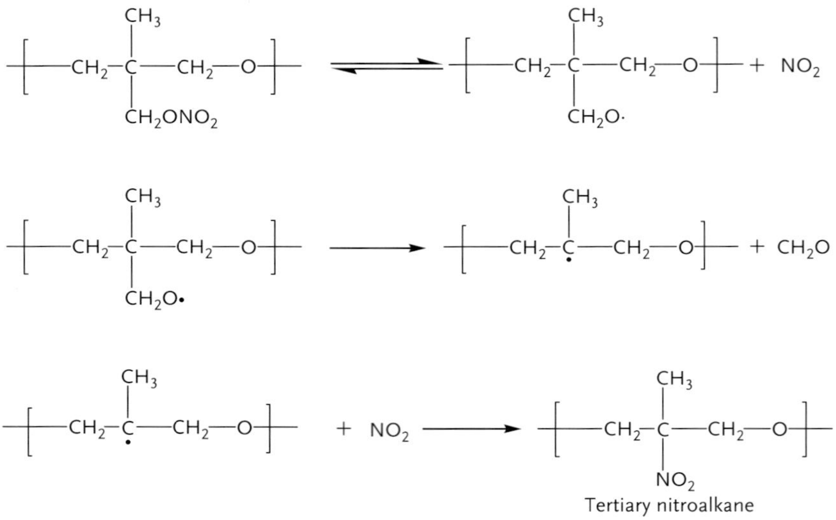

Scheme 3.14 Formation of nitroalkane from decomposition of Poly(NIMMO) [39].

decomposition. Thus, NC-based formulations should be capable of retaining better mechanical properties under high-temperature ageing as compared with those of the Poly(NIMMO) ones. However, NC-based propellant/explosive formulations should always be mixed with stabilizers to avoid the explosive autocatalytic decomposition of NC, which is not required for the Poly(NIMMO) binder formulations.

3.3.3
Poly(Glycidyl Nitrate) (PGN)

PGN is another important member of nitrate ester group of binders that has a high oxygen balance and density. Thermal decomposition of PGN has been studied by TGA [35(a)] and a variety of calorimetric techniques [40].

Arrhenius plots generated for PGN and Poly(NIMMO) from TGA analysis are compared in Figure 3.5 [35(a)]. PGN, similar to Poly(NIMMO), follows a first-order kinetics throughout the initial 50% weight loss (144–180 °C) without undergoing autocatalysis. The calculated kinetic parameters for thermal decomposition of PGN are: $E_A = 193.43$ kJ/mol and $\ln A = 44.4$ (1/s), which are comparable to those for Poly(NIMMO). The magnitude of activation energy suggests that the decomposition proceeds by the scission of the $-O-NO_2$ bond in the polymer.

Bunyan [40] calculated the activation energy for thermal decomposition of PGN at moderate temperatures (65–81 °C) using heat flow calorimetry, and found that the values are significantly lower (95.6–105.4 kJ/mol). This result suggests possible changes in the decomposition mechanism as room temperature is approached.

At faster heating rates (>100 °C/s), similar to the situation with NC and Poly(NIMMO), the rate law linearizes with a factor of $n = 2$, which does not necessarily

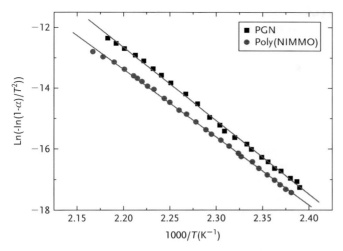

Figure 3.5 Comparison of Arrhenius plots for PGN and Poly(NIMMO). Values for the plot taken with permission of Elsevier from Ref. [35(a)].

mean that the reaction obeys a second-order rate law. The activation energy parameters for fast rate thermal decomposition of PGN are: $E_A = 130.21 \pm 2.1$ kJ/mol and $\ln A = 33.85$ (1/s) [38(b)].

The de-curing problems of isocyanate cured PGN binders have been discussed in detail in Section 3.2.2.1. End-modification of the prepolymer chain by hydroxyl groups was suggested as a promising route to overcome the problem of instability [41]. Stoltz and Peiris [42] applied T-jump and FTIR spectroscopy to understand the effects of hydroxyl end-modification on the decomposition of the PGN prepolymer. Activation energies were calculated relative to the formation of CH_2O and CO_2 gases formed during the decomposition process by applying zero-order kinetics. Activation energies derived on the basis of CO_2 formation (92 kJ/mol for unmodified PGN polymer and 75 kJ/mol for hydroxyl end-modified) were erroneous, because CO_2 formation is not specific to the initial decomposition step of the polymer. On the other hand, activation energy calculated relative to CH_2O formation (134 kJ/mol) was comparable to the values calculated from DSC measurements (145.28 kJ/mol), implying that CH_2O formation is the probable initial step in the decomposition of PGN (Scheme 3.15). Ling and Wight have also proposed a similar mechanism for PGN decomposition based on the results of laser photolysis experiments [43]. Thus, thermal decomposition kinetics of PGN prepolymer is not affected by end-modification.

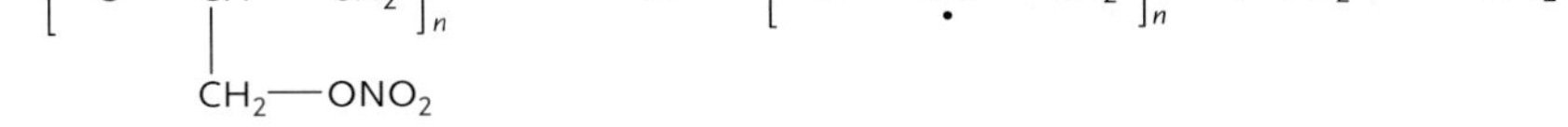

Scheme 3.15 Initial thermal decomposition step of PGN.

3.4
Combustion of Nitrate Ester Polymers and Propellants

3.4.1
NC Based Double-Base Propellants

NC based double-base (DB) propellants are the most popular among the nitrate ester polymer propellants and they have been investigated for a long time, starting in the 1950s [44]. A schematic diagram of the combustion process of the DB propellant is illustrated in Figure 3.6.

The propellant begins to react about 0.3–0.05 mm below the surface, where the temperature becomes high enough for the scission of the –O–NO_2 bond, forming NO_2 and aldehydes [45]. This region is known as the foam zone, where the solid becomes a viscous liquid and bubbles form, which give the region its name [44]. At low pressures (under 100 atm), the primary flame (fizz zone) and secondary flame (luminuous flame) are separated by a dark zone. The primary flame zone involves the exothermic reaction between NO_2 and aldehydes to form NO, CO, CO_2, and H_2O, resulting in a steep temperature rise (from 700 to 1700 K) [44, 46]. After the fizz reaction, the temperature levels off in the dark zone [45]. The reaction at the secondary luminuous zone is the reaction between NO and hydrocarbons to form the final combustion products (N_2, CO, and H_2O) along with the luminous flame having a temperature in the range of (2100–3100 K) [44, 45].

At low-pressure ranges, the secondary flame is too far away to have any effect on the burning surface. The burning rate is then entirely under the influence of the primary flame and the pressure exponent through the usual empirical law ($r_b = ap^n$) with $n \leq 0.7$.

As the pressure increases, the secondary flame merges with the primary and the width of the dark zone decreases. Thereafter, the secondary flame is considered to be driving the degradation reaction in the condensed phase, which is reflected as a change in slope of the burn-rate profile (Figure 3.7) [44].

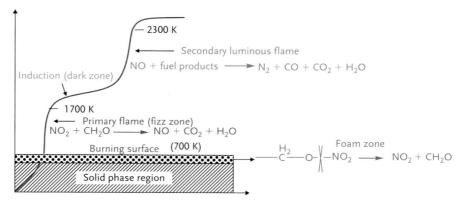

Figure 3.6 Schematic of a combustion process of NC propellant showing various zones.

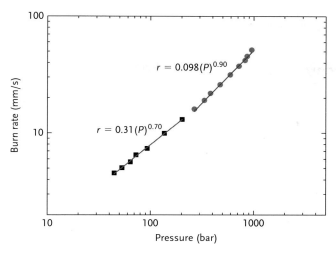

Figure 3.7 Burning rate versus pressure profile for a DB propellant. Values for the plot taken with permission of Wiley from Ref. [44].

It is interesting to contrast the combustion behavior of nitrate ester polymers with that of the azido polymers discussed in the previous chapter. Combustion of nitrate ester polymers is controlled by the heat feedback mechanism of gas-phase reactions to the propellant surface. Conversely, combustion behavior of azido polymers is controlled by the highly exothermic decomposition of the azido groups in the condensed phase. Hence, the azido polymers exhibit higher burn rate and have lower pressure sensitivity as compared with that for nitrate ester polymers.

3.4.2
Composite Modified Double-Base (CMDB) Propellants

Double-base propellants exhibit relatively low burning rates of $r = 10–25$ mm/s at 7 MPa with a pressure exponent of $n = 0–0.3$ [47]. Even though DB propellants provide a wide range of burn rates at low-pressure exponents with a non-smoky exhaust, they have lower performance and density compared with the composite propellants [48]. In order to improve the performance of DB propellants, they are mixed with high-energy materials to obtain composite modified double base (CMDB) propellants. They are attractive in various applications due to the flexibility in burn rate, superior mechanical properties, high performance, and non-smoky exhaust [49].

The combustion mechanism of CMDB propellants is fundamentally different from that of the DB, which has homogenous combustion zones. The combustion zone of CMDB is non-homogenous due to the heterogeneous nature of the propellant. The crystalline oxidizers interact with the surrounding DB matrix to produce multiple flamelets at the burning surface, and the burning rate behavior is dependent on the physical and chemical modes of the energetic oxidizer [50].

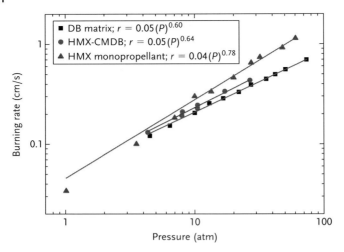

Figure 3.8 Burning rate profiles of HMX particles in DB matrix. Profiles of DB matrix and HMX monopropellant are also plotted for comparison. Data for the plot taken with permission of Elsevier from Ref. [50, 54].

Among the energetic oxidizers, the addition of cyclic nitramines, such as Royal Demolition Explosive (RDX) and Her Majesty's Explosive (HMX) in the DB matrix, is expected to improve not only the energy output but also the thermal stability of propellants. The major contribution to the energy output of this class of propellants is attributed to the positive heat of formation of RDX and HMX and formation of low molecular weight gases [51]. However, the major problems associated with these propellants are the low burn rates and high pressure exponents [50].

Specifically, it has been reported that the burn rate of the HMX based doublebase propellants decreases with increasing concentration of HMX. This is due to the fact that energy release in the fizz zone just above the burning surface is considerably reduced because of the endothermic melting of HMX [51] and its combustion away from the surface in the luminous zone of the double-base flame [52]. Moreover, the kinetics of energy release from the fizz zone are also slowed down by the shift in the equivalence ratio of NO_2 : aldehydes towards a more fuel-rich composition of HMX [53].

As illustrated in Figure 3.8, the burning rate of HMX particles embedded in a DB matrix did not appear to differ significantly from that of the DB matrix itself and it is lower than the burn rate of HMX monopropellant alone. The flame structure is similar to that of the DB propellant compositions [52].

Incorporation of ammonium perchlorate (AP) instead of HMX results in a different burning rate profile as illustrated in Figure 3.9.

The burn rate of AP/CMDB is remarkably higher than that of DB matrix and AP monopropellant. This is attributed to the fact that the interaction between oxidizer-rich AP decomposition products and fuel-rich DB matrix decomposition products leads to an increase in the reaction rate in the fizz zone, which contributes to elevation of the condensed-phase temperature [52] and hence the burn rate is

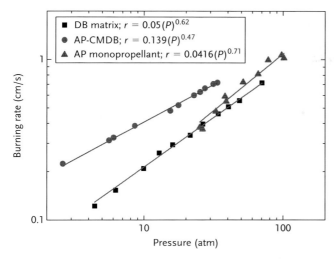

Figure 3.9 Burning rate profiles for AP particles embedded in DB matrix. Profiles of DB matrix and AP monopropellant are plotted for comparison. Data for the plot taken with permission of Elsevier from Ref. [50,54].

increased. Furthermore, in contrast to the HMX compositions, the AP composition gave an intense flame with a reduction in the dark zone, leading to greater feedback from the gas phase to the burning surface [50, 52].

Even though AP/CMDBs provide better burn rates, they suffer from several disadvantages, the most critical being combustion instability [55]. Addition of Al to AP-CMDB composition is found to mitigate the combustion instability problem by the formation of a thin layer of molten Al metal protected with alumina on the burning surface [52]. However, Al addition does not contribute to the burn rate of the composition, because combustion of Al involves the breaking of the oxide layer at 2500 K in the high-temperature zone [52]. AP/Al-based CMDB propellants have a theoretical specific impulse of 260–265 s and burn rates of 11–16.5 mm/s at 6.8 MPa with a pressure index of 0.34 [56].

Another problem is the catalytic effect of AP on the decomposition of nitroglycerin (NG) resulting in autocatalysis. The strong autocatalytic effect is induced by the mutual interaction of the decomposition products of the DB matrix and AP particles [57], particularly perchloric acid [58]. Several stabilizer systems, such as resorcinol, alkoxy-phenoxy-benzenes, metal oxides, molecular sieves [59], and a dual stabilizer system comprising *N*-methyl-*p*-nitroaniline and aluminum silicate molecular sieves [60] have been recommended to curtail the problem of instability of AP/CMDB compositions. The extended anion network present in the molecular sieve based stabilizers helps to effectively remove elements such as nitric oxides and acids responsible for the autocatalytic effect through acid–base interactions [59].

AP-based compositions produce secondary smoke consisting of chlorine containing compounds that pose pollution hazards. Hence, research interest has been shifted to nitramine-based CMDB propellants, which are marginally superior to

AP-based compositions in energetics and offer cleaner burning characteristics [61]. However, nitramine-based CMDB propellants suffer from lower burn rates due to a decrease in heat feedback to the burning surface. Such compositions also resist ballistic modification, because many of the catalysts do not have much effect on the rate-controlling step of their combustion [61].

Intensive research is ongoing all over the world to augment the ballistics of nitramine-CMDB compositions by addition of metals. Zirconium (Zr) based formulations gave superior burn rates (of the order of 7–14 mm/s in the pressure range of 4.9–10.8 MPa) as compared with nickel (Ni), titanium (Ti) or aluminum (Al) compositions (Table 3.4) [61]. Unlike Al-, Ni-, or Ti-based formulations, Zr-based RDX/CMDB compositions exhibit stable combustion at lower pressure [61].

Among the ballistic modifiers, a basic lead salicylate + cuprous oxide + C-black combination and copper chromite provides superior catalytic effect to burn-rate enhancement for the RDX/Zr/CMDB propellant systems compared with that of ferric oxide and ferric acetyl acetonate [61].

Lately, hexanitrohexaazaisowurtzitane (CL-20) has emerged as a high-performance alternative to RDX and HMX in CMDB propellants [62, 63]. It has been observed that CL-20 based aluminized CMDB propellants offer superior burn rates of the order of 6.5–9.7 mm/s in the pressure region of 5–9 MPa as compared with RDX incorporated propellants [63] (Figure 3.10). The CL-20 composition also provides a lower pressure index (0.33) as compared with the RDX-based compositions (0.49).

As illustrated in Figure 3.11, addition of copper chromite (CC) as catalyst to CL-20/Al/CMDB propellant improves the burn rate (4.3–13.6 mm/s in the pressure range of 2–9 MPa) and also extends the low pressure combustion limit. Furthermore, substituting the inert DEP plasticizer with energetic GAP or BDNPF/A enhances the burn rate of CC modified CL-20/Al/CMDB propellants by 45–170% [63].

Ammonium dinitramide (ADN) and hydrazinium nitroformate (HNF) are emerging as potential eco-friendly oxidizers for CMDB propellants and appear to be superior to RDX and HMX [64]. HNF has certain advantages over ADN, such as the simple method of synthesis, non-hygroscopic nature, higher density, and melting-point [64, 65]. Theoretical performance predictions of CMDB propellants with AP, ADN, and HNF oxidizers are compared in Table 3.5.

Table 3.4 Variation of burn rates with respect to pressure and respective pressure indices for RDX/CMDB propellants containing different metals. Values are taken from [61].

Metal content	4.9 (MPa)	6.8 (MPa)	8.8 (MPa)	10.8 (MPa)	Pressure index
Control (NC/NG/RDX)	–	6.4	8.7	10.3	1.02
+10% Al	–	6.9	8.4	10.6	0.92
+10% Ni	–	6.6	7.9	8.9	0.65
+10% Ti	–	7.7	9.3	10.9	0.75
+10% Zr	5.6	7.4	8.9	11.5	0.88

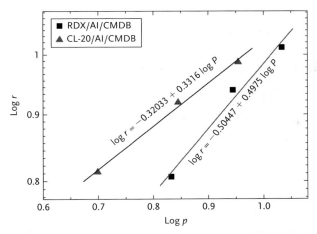

Figure 3.10 Comparison of burn rates and pressure exponents for CL-20 and RDX based aluminized CMDB propellants. Values for the plot are taken from Refs. [61] and [63].

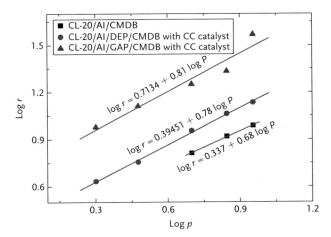

Figure 3.11 Burn-rate profiles for CL-20/Al/CMDB systems with CC catalyst and energetic GAP co-plasticizer. Values for the plot are taken from Ref. [63].

Table 3.5 Theoretical performance prediction of CMDB propellants with different oxidizers [65].

CMDB propellant with oxidizer	Flame temperature (°C)	Molecular weight of exhaust gases	Specific impulse (s)
AP	3646	31.29	260.8
ADN	3634	30.17	263.6
HNF	3661	30.08	264.8

As illustrated in Table 3.5, the HNF formulation is capable of producing higher specific impulses with higher burning rates [66] than AP- or ADN-based CMDB formulations.

3.4.3
Poly(NIMMO)-Based Composite Propellants

The energetics and ballistics of traditional double or triple base propellants becomes impressive by the loading of a DB matrix with high energy particulate explosive filler materials, such as the sensitive cyclic nitramines RDX and HMX. However, this has led to increasing the vulnerability of propellants to hazardous stimuli. A prime example of this scenario is the violent increase in response when the double JA2 propellant is loaded with RDX to produce the JAX propellant [67].

The use of Poly(NIMMO) as a rubbery energetic binder has opened up the possibility of achieving high impetus levels whilst attaining the necessary low-vulnerability ammunition (LOVA) characteristics [68]. As the composition contains an energetic binder prepolymer, a high performance could be maintained by increasing the proportion of the energetic binder whilst reducing the loading of the sensitive explosive filler [68(b)].

It can be observed from Table 3.6 that Poly(NIMMO)-based propellant formulations offer higher impetus and relatively lower flame temperatures, which are at least 300 K below that of the traditional non-LOVA JA2 propellant having equivalent energy.

New propellants based on clean energetic oxidizers such as HNF and ADN with Poly(NIMMO) and PGN have been projected as futuristic propellants, due to (i) much higher specific impulse than any of the existing propellants and (ii) chlorine free exhaust products [64]. As illustrated in Figure 3.12, the specific impulse of the propellants could be further enhanced by the addition of aluminum metal [64].

It is well known that HNF propellants have high pressure exponents [70]. Several burn-rate modifiers have been tested with Poly(NIMMO)/HNF propellants to modify the burn rates and tune the ballistics at the TNO laboratories in The Netherlands [71]. The chemical nature of burn-rate modifiers has not been divulged in Ref. [71]. The burn rates were reasonably varied with the addition of the burn-rate modifiers to Poly(NIMMO)/HNF based compositions, but they were higher than what would be desirable for a solid propellant composition [71].

Table 3.6 Comparison of ballistic data of Poly(NIMMO) gun propellant formulations with that of the NC based JA2 propellant [68].

Propellant	Impetus (kJ/kg)	Flame temperature (K)	Reference
LOVA 4 (Poly(NIMMO)/HMX)	1228–1230	3037	[68(a)]
JA2 (NC based)	1139	3390	[69]

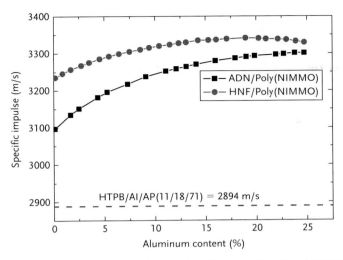

Figure 3.12 Vacuum specific impulse for ADN and HNF with Poly(NIMMO) binder with respect to aluminum content. The specific impulse of HTPB/Al/AP propellant is provided as a baseline for comparison. Figure plotted with values from Ref. [64].

One of the attractive features of the Poly(NIMMO)/HNF propellants is its low burn-rate sensitivity towards temperature (of the order of 0.55×10^{-3}/K), which is much lower than that of conventional composite or double-base propellants and neat HNF [71]. The Poly(NIMMO)/HNF propellants have comparable rotter impact sensitivity with non-alumnized AP/HTPB propellant but are more sensitive than Al/AP/HTPB propellants. HNF/Al/Poly(NIMMO) propellants have substantially lower friction sensitivity as compared with pure HNF and are of the same order or slightly less than a non-aluminized AP/HTPB propellant [71].

As part of the collaborative program between Sweden and the United Kingdom, extensive studies were conducted on various ADN propellant formulations with energetic binders (Poly(NIMMO), PGN, and GAP) for their chemical compatibility, chemical stability, cure characteristics, and burn rates [72]. Among the energetic polymers, cured Poly(NIMMO) was observed to be compatible with ADN. On the other hand, PGN was significantly incompatible, associated with the modified 1,2-dihydroxy end group present on the PGN prepolymer [72]. All the isocyanate curing agents show marked incompatibility with ADN, among which H_{12}MDI (4,4'-dicyclohexylmethane diisocyanate) is the most appropriate, which shows the least reactivity towards ADN. Curing studies on Poly(NIMMO)/ADN/dioctyl sebacate (DOS)/H_{12}MDI formulation indicate that ADN interferes with the curing reaction through its reaction with the isocyanates. However, the tensile properties and overall stability of the cured propellant show excellent results [72].

Results of burn-rate measurements illustrated in Figure 3.13 show that the burn rates of Poly(NIMMO)/ADN propellants are in the range of moderate burning AP

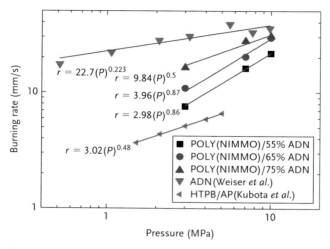

Figure 3.13 Burn-rate profiles for Poly(NIMMO)/ADN propellants. Burning rate profiles of ADN and a conventional HTPB/AP propellant are also plotted for comparison. Values for the plot are taken from Refs. [72–74]. The burn rate values of Poly(NIMMO)/ADN propellants are reproduced with the permission of Niklas Wingborg and Dr. Carina Eldsater, FOI, Sweden.

composite propellants [72, 74]. The burn rates increases with the amount of ADN in the propellant. All the propellants have lower burn rates than the pure ADN, which shows a mesa/plateau effect in the range of 4–7 MPa [73]. However, all the ADN propellants, excluding the one containing 75% ADN, show higher than desirable burn-rate exponents, which is a problem found with a number of advanced propellant formulations. Similar to the situation with HNF propellants, suitable burn-rate modifiers should be evaluated with ADN propellants, which may reduce the exponents to optimal values [72].

The small scale hazard test data for Poly(NIMMO)/ADN propellant formulations are presented in Table 3.7.

The small scale hazard test indicates that the Poly(NIMMO)/ADN formulation is less impact sensitive as compared with the NC-based formulation. However, the

Table 3.7 Comparison of small scale hazard test data for Poly(NIMMO)/ADN propellant formulation with that of NC-based propellant formulation [72, 75].

Test	Poly(NIMMO)/ADN	NC-based propellant
Rotter impact test, F of I (figure of insensitivity)	37	21.8
Friction (rotary friction, F of F)	2.0	Friction insensitive up to 14 kg
Temperature of ignition	145 °C	160 °C
Electrostatic discharge	No ignition at 4.5 J	No reaction for 20 trials at 0.75 J

ADN formulations suffer from high friction sensitivity and have a low temperature of ignition. Even though the Poly(NIMMO)/ADN formulation is predicted to offer enhanced performance because of higher burn rates, the problems of friction sensitivity, low ignition temperature, and higher burn rate exponents should be circumvented in order to put the formulation into operational use.

3.4.4
PGN-Based Composite Propellants

PGN is one of the earliest energetic nitropolymers to be investigated as a binder for composite propellants. PGN and PGN propellants were examined at the Jet Propulsion Laboratory by Ingnam and Nichols and at Aerojet General Corporation by Shookhoff and Klotz in late 1950s [76]. However, the major impediment to its further development was the tedious/hazardous synthetic route of monomer preparation and the inability to polymerize the monomer to obtain a polymer with functionality greater than two. With the identification of safer synthetic routes and alterations to the polymerization conditions in the early 1990s, PGN started to be used widely as a binder in composite propellants. Table 3.8 compares the performance and density of PGN propellants for large launch vehicles with that of the present standard space shuttle propellants consisting of HTPB or PBAN binders [77].

As illustrated in Table 3.8, PGN based propellant formulations provide almost similar specific impulses as the HTPB/AP/Al formulation, but at low oxidizer content. The low oxidizer content renders the PGN formulations more insensitive towards external stimuli and also helps in the processing. Furthermore, the PGN formulations employ ammonium nitrate (AN) as the oxidizer, which will not produce the corrosive and environment polluting HCl in their exhaust, in contrast to the situation with ammonium perchlorate (AP) oxidizer. Thus, the AN/PGN system incorporates the potential for high energetic non-polluting propellants, which are really less sensitive and exhibit smokeless burning.

However, the low oxidizing ability of AN limits the performance of propellant formulations, which is augmented by adding high-performance nitramines (CL-20 and HMX) and energetic plasticizers into the formulation [78].

It is clear from Figure 3.14 that addition of nitramines significantly enhances the performance of the AN/PGN formulation. At least 10 wt% of nitramine

Table 3.8 Theoretical specific impulses and densities for PGN based propellant formulations as compared with the conventional HTPB based propellant formulations [77].

Property	PGN/AN/Al (30 : 50 : 20)	PGN/AN/HMX/Al (30 : 48 : 22) (AN : HMX in 3 : 1 ratio)	HTPB/AP/Al (12 : 68 : 20)
Specific impulse (I_{sp}) (s)	259.6	264.0	265.3
Density	1.74	1.77	1.88

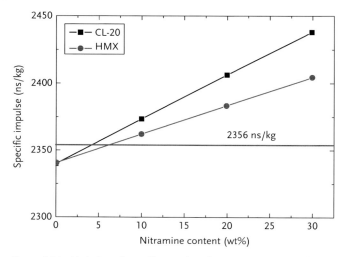

Figure 3.14 Variation of specific impulse of PGN/AN/energetic plasticizer propellants with respect to the addition of different nitramines to the formulation. The energetic plasticizer used in the formulation is a mixture of TMETN and GAPA in 4 : 1 ratio. Values for the plot are taken from [78].

should be added to the formulation to cross over the minimum performance requirement of 2356 ns/kg. Among the nitramines compared, the more powerful CL-20 containing compositions are able to deliver higher impulses as compared with that containing HMX.

Triaminoguanidine nitrate (TAGN) and guanidine-5,5′-azotetrazolate (GZT) were found to be efficient burn-rate modifiers for the PGN/AN/nitramine/ energetic plasticizer compositions [78]. The burning rate profiles of PGN/AN/ nitramine formulations with or without burn-rate modifiers are presented in Figure 3.15.

It could be observed that all the PGN formulations exhibit higher burn rates as compared with the HTPB/AP/Al formulations especially at higher pressures. The burn rates of PGN/AN/CL-20 formulations are enhanced by the addition of burn-rate modifiers to the formulation. The burn rate of PGN/AN/CL-20 formulation at 7 MPa increases from $r=6$ mm/s ($n=0.63$) to $r=11.2$ mm/s ($n=0.5$) with the addition of 20% of TAGN to the formulation. TAGN is also sufficiently instrumental in reducing the pressure exponent to desirable levels of $n \leq 0.5$. Even though addition of GZT also improves the burn rate ($r=7.8$ mm/s at 7 MPa), it does not have much effect on the pressure exponent value ($n=0.58$) [78]. The AN/CL-20/PGN compositions containing TAGN and GZT deliver an average specific impulse of 241 s, which is a type of limit of performance for less sensitive minimum smoke propellants based on AN and PGN. The specific impulse of TAGN formulations could be made even higher than that of the traditional HTPB/AP/Al composite propellant by the addition of 20% Al to the composition [79].

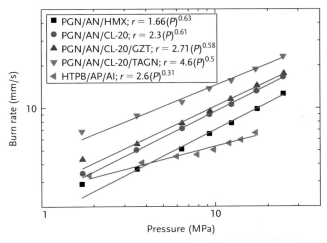

Figure 3.15 Burn-rate profiles of different PGN/AN/nitramine composition with or without burn-rate modifiers with the corresponding pressure exponents. Burn-rate profile of HTPB/AP/Al propellant is also plotted for comparison. Values for the plots are taken with permission of Wiley-VCH from Refs. [74] (HTPB/AP/Al) and [78] (PGN/AN/nitramine formulations.

Chemical stability of PGN/AN/CL-20 compositions as determined by microcalorimetry indicates that the stability of formulations containing TAGN is not sufficient (in accordance with the thermal instability of nitrate esters), but that with GZT is tolerable for propellant applications [78]. Lead citrate provides good chemical stability with 20–30% TAGN in short-term tests [79]. The problematic long- term stability of AN/TAGN formulations and the high pressure exponent values of AN/GZT formulations are two major hindrances facing the full-scale development of these propellant formulations. Once these problems are overcome the AN/TAGN formulations with PGN will play an important role in the field of minimum smoke propellants and also in the metallized form.

3.5
Nitropolymer-Based Explosive Compositions

PBX-9404 is a pressed plastic bonded explosive using NC binder. It has the formulation of 94 wt% HMX, 3 wt%NC, and 3 wt% CEF (tris(β-chloroethyl) phosphate) [80]. The performance properties of PBX-9404 are compared with well known melt cast explosives (Octol and Comp B) in Table 3.9.

Clearly, PBX-9404 is the explosive with the best performance parameters (higher detonation velocity and detonation pressure) among the group. However, as shown in Table 3.10 it is too sensitive for military applications and also lacks the thermal stability for use in high-temperature environments [80, 81].

Table 3.9 Performance of PBX-9404 compared with Comp B and Octol [81].

Explosive	Approximate composition	Initial density (g/cm^3)	Detonation velocity (m/s)	Detonation pressure (GPa)
Comp B	60 : 40 RDX/TNT	1.65	7800	24.4
Octol	76 : 24 HMX/TNT	1.81	8476	34.3
PBX-9404	94 : 3 : 3 HMX/NC/CEF	1.84	8800	36.5

Table 3.10 Sensitivity parameters of PBX-9404 compared with Comp B [81].

Explosive	Impact sensitivity LLNL 12B (cm)	Shock sensitivity (small scale gap test) (mm)	Thermal stability DTA exotherm (onset) (°C)
PBX-9404	40	2.46	180
Comp B	60	0.41–0.66	214

PBX-9404 is largely being replaced by Estane based PBX-9501 (95 : 2.5 : 2.5 HMX/Estane/BDNP-F) composition.

Experimental Poly(NIMMO) based high performance PBX compositions have been developed for metal accelerating applications that possess improved performance over widely used PBXN-110 (88% HMX/12% HTPB) [82] and have an equivalent insensitive munition (IM) response. The explosive filler of the PBX is HMX [83]. The performance properties of the PBX compositions were studied with respect to the addition of three different energetic plasticizers: butyl NENA, K10, and GAPA [82, 84]. Performance modeling suggests both the velocity of detonation and detonation pressure increase with the energetic plasticizer content [84].

With respect to the energetic plasticizer, the predicted performance parameters follow the following trend [82]:

- velocity of detonation: butyl NENA > GAPA > K10;
- P_{CJ}: butyl NENA > GAPA = K10.

Both velocity of detonation and P_{CJ} (pressure at the Chapman–Jouguet point) are highest for the formulation containing butyl NENA plasticizer (RF-67-43).

As shown in Table 3.11, the mean velocity of detonation (measured by triggering the ionization probes) for RF-67-43 was 1% above PBXN-110, and the predicted detonation pressure was 5.8% above PBXN-110. The shock sensitivities are on par with that for PBXN-110. The noticeable feature is the higher density and lower total solid content for RF-67-43, which will be helpful in miniaturization and easy processing of the PBX formulation.

PBX formulations based on the Poly(NIMMO) binder have been developed with the insensitive explosive 3-nitro-1,2,4-triazole-5-one (NTO) and HMX designed for insensitive munitions. The energetic binder compensates for the small loss in performance due to the replacement of more energetic HMX with less sensitive

Table 3.11 Comparison of performance and shock sensitivity parameters of RF-67-43 explosive with PBXN-110 [82].

Formulation	RF-67-43	PBXN-110
Explosive	HMX	HMX
Binder	Poly(NIMMO)	HTPB
Plasticizer	Butyl NENA	IDP (isodecyl pelargonate)
Total solids (%)	84 (w/w)	88 (w/w)
Density (g/cm^3)	1.74	1.678
Velocity of detonation (km/s) (measured using ionization probe)	8.47	8.39
P_{CJ} (GPa)	32.9	29
Large scale gap test (cards) (MIL-STD-1751A) Method 1041(NOL)	162	154–178

NTO. The United Kingdom has developed experimental PBX formulations using Poly(NIMMO)/Al/HMX/NTO (CPX series) directed towards underwater blast charge applications [83]. The formulations and velocity of detonation data of CPX series formulations are presented in Table 3.12.

Extended large scale gap tests on CPX formulations show all of them to be less shock sensitive than the HMX reference, ORA 86 (86% HMX and 14% inert polyurethane binder) [83] (Table 3.13).

The composition CPX 413 has passed the HD 1.6 test series and is ranked as an extremely insensitive detonating substance (EIDS) [84].

New explosive molecules such as 1,1-diamino-2,2-dinitroethylene (FOX-7) with inherent lower sensitivity provide safe handling of ammunition. Different PBX formulations based on FOX-7 and energetic binders Bu-NENA plasticizer and

Table 3.12 Formulations and velocities of detonation for the CPX series of explosives based on Poly(NIMMO). The parameters for HTPB based HX 310 are also given for comparison [83].

Composition (country)	Explosive		Metal fuel	Binder (%)	Plasticizer (%)	Density (g/cm^3)	VOD (m/s)
	NTO (%)	HMX (%)					
CPX 412 (UK)	50	30	–	Poly(NIMMO); 10	K10; 10	1.66	7200
CPX 413 (UK)	45	35		Poly(NIMMO); 10	K10; 10	1.74	8150
CPX 450 (UK)	40	20	20	Poly(NIMMO); 10	K10; 10	1.85	7762
CPX 458 (UK)	30	30	20	Poly(NIMMO); 10	K10; 10	1.86	7761
HX 310 (UK)	25	47	NG:10	HTPB; 18	–	1.57	7750

Table 3.13 Extended large scale gap test results for selected CPX series explosives [83].

Test formulation	Thickness of barrier (mm)
CPX 450	26
CPX 413	33
ORA 86	80

H_{12}MDI curing agent have been prepared at the Swedish Research Agency (FOI) with the objective of obtaining a FOX-7 formulation with similar performance (in terms of detonation velocity) as that of Comp B [85] but with improved sensitivity. Computed detonation velocity values show that among the energetic binders, PGN provides the best performance followed by GAP and AMMO (3-azidomethyl 3-methyl oxetane)/BAMO (3,3-bis(azidomethyl) oxetane) [85]. As shown in Figure 3.16, energetic binder formulation PBX (FOF3) containing only FOX-7 as the filler is unable to achieve the performance of Comp B. When HMX was added along with FOX-7 in the formulation in order to trade-off safety for performance, the energetic binder PBX formulation (FOF5) nearly achieves the performance target. It is noteworthy that FOF5 could achieve the performance of Comp B at a lower loading of sensitive HMX filler as compared with the inert binder PBX formulation (QRX077).

Therefore, energetic binder PBX formulations provide improved insensitivity without compromising the performance.

The response of PGN/HMX/FOX-7 formulations towards IM tests are compared with that of Comp B in Table 3.14.

PGN/FOX-7/HMX formulation shows lower response type to slow cook-off, fast cook-off, and bullet impact tests as compared with Comp B. The formulation

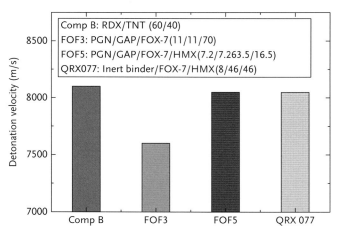

Figure 3.16 Comparison of theoretical performance of different PBX formulations containing FOX-7 and HMX explosives with Comp B [85–87].

Table 3.14 Comparison of sensitivity parameters of PGN/HMX/FOX-7 formulation with Comp B [85].

Formulation	Response to slow cook-off	Response to fast cook-off	Bullet impact
PGN/HMX/FOX-7	Fire/deflagration	Deflagration	Fire
Comp B (RDX/TNT)	Detonation	Detonation	Detonation

did not detonate during a detonation test in steel tubes and burned without damage to the surroundings in a slow cook-off test [85]. Thus, PGN/HMX/FOX-7 formulations have a promising future for use in insensitive munitions and as a possible replacement for Comp B.

One word of caution here is that PGN/FOX-7 formulations are more impact sensitive than the ingredients constituting the PBX [85(a)]. This problem should be solved before commercializing the PGN/FOX-7 formulations as insensitive munitions.

References

1 Davis, T.L. (1972) *The Chemistry of Powder and Explosives*, Angriff Press.
2 Blaker, R.H., Badger, R.M., and Noyes, R.M. (1947) Molecular properties of nitrocellulose. II Studies of molecular heterogeneity. *J. Phys. Chem.*, **51** (2), 574–579.
3 Taylor, J., Hall, C.R.L., and Thomas, H. (1947) The thermochemistry of propellant explosives. *J. Phys. Chem.*, **51** (2), 580–592.
4 Taylor, J. and Hall, C.R.L. (1947) Determination of the heat of combustion of nitroglycerine and the thermochemical constants of nitrocellulose. *J. Phys. Chem.*, **51** (2), 593–611.
5 Jessup, R.R. and Prosen, E.J. (1950) Heats of combustion and formation of cellulose and nitrocellulose. *J. Res. Natl. Bur. Stand.*, **44**, 387–393.
6 Barbosa, I.V.M., Merquior, D.M., and Peixoto, F.C. (2006) Estimation of kinetic and mass-transfer parameters for cellulose nitration. *AIChE. J.*, **52**, 3549–3554.
7 (a) Murbach, W.J., Fish, W.R., and Van Dolah, R.W. (1953) Polyglycidyl nitrate. Part 1 Preparation and characterization of glycidyl nitrate, NAVORD report 2028, NOTS 685, US Naval Ordnance; (b) Meitner, J.G., Thelen, C.J., Murbach, W.J., and Van Dolah, R.W. (1953) Polyglycidyl nitrate. Part 1 Preparation and characterization of glycidyl nitrate, NAVORD report 2028, NOTS 686, US Naval Ordnance.
8 (a) Norman, P.C. and Millar, R.W. (1992) Preparation of nitratoalkyl substituted cyclic esters. US Patent 5,145,974; (b) Millar, R.W., Colclough, M.E., Golding, P., Honey, P.J., Paul, N.C., Sanderson, A.J., Stewart, M.J., Volk, F., and Thompson, B.J. (1992) New synthetic routes for energetic materials using dinitrogen pentoxide. *Phil. Trans. Phys. Sci. Eng.*, **339**, 305–319.
9 Colclough, M.E., Desai, H., Millar, R.W., Paul, N.C., Stewart, N.J., and Golding, P. (1993) Energetic polymers as binders in propellants and explosives. *Polym. Adv. Tech.*, **5**, 554.
10 (a) Highsmith, T.K., Sanderson, A.J., Cannizo, L.F., and Hajik, R.M. (2002) Polymerization of PGN from high purity GN synthesized from glycerol. US

Patent 6,362,311; (b) Paraskos, A.J., Sanderson, A.J., and Cannizo, L.F. (2004) Polymerization of GN via catalysis with BF$_3$ THF: Compatibility with activated monomer mechanism. November 2004 IM/EM Symposium, San Francisco; (c) Cannizzo, L.F., Hajik, R.M., Highsmith, T.K., Sanderson, A.J., Martins, L.J., and Wardle, R.B. (2000) A new low cost synthesis of PGN. 31st International Annual Conference of ICT, 36/1–36/9.

11 Bednarek, M., Kubisa, P., and Penczek, S. (2001) Multihydroxyl branched polyethers. 2. Mechanistic aspects of cationic polymerization of 3-ethyl-3-(hydroxymethyl) oxetane. *Macromolecules*, **34**, 5112–5119.

12 Kubisa, P. and Penczek, S. (1999) Cationic activated monomer polymerization of heterocyclic monomers. *Prog. Polym. Sci.*, **24**, 1409–1437.

13 Kubisa, P. (2003) Hyperbranched polyethers by ring-opening polymerization: contribution of activated monomer mechanism. *J. Polym. Sci. A: Polym. Chem.*, **41**, 457–468.

14 (a) Miller, R.L., Day, R.S., and Stern, A.G. (1996) Process of producing improved PGN. European Patent EP 0471489B1; (b) Miller, R.L., Day, R.S., and Stern, A.G. (1992) Process of producing improved PGN. US Patent 5,120,827.

15 Stewart, M.J. (1994) Polymerization of cyclic ethers. US Patent 5,313,000.

16 Ingham, J.D. and Nichols, P.L. Jr (1959) High performance PGN-polyurethane propellants, Publication Number 93, Jet Propulsion Laboratory.

17 Desai, H.J., Cunliffe, A.V., Lewis, T., Millar, R.W., Paul, N.C., Stewart, M.J., and Amass, A.J. (1996) Synthesis of narrow molecular weight hydroxy telechelic polyglycidyl nitrate and estimation of theoretical heat of explosion. *Polymer*, **37** (15), 3471–3476.

18 Willer, R.L., Stern, A.G., and Day, R.S. (1993) Isotactic poly (glycidyl nitrate) and syntheses thereof. US Patent 5,264,596.

19 Sanderson, A.J., Martins, L.J., and Dewey, M.A. (2005) Process for making stable cured PGN and energetic compositions comprising same. US Patent 6,861,501 B1.

20 (a) Kim, J.S., Cho, J.R., Lee, K.D., and Kim, J.K. (2007) 2-Nitroethyl oxirane, poly (2-nitratoethyl oxirane) and preparation method thereof. US Patent 7,288,681 B2; (b) Kim, J.K., Kim, J.S., Cho, J.S., and Kim, J.K. (2003) A new energetic prepolymer–structurally stable PGN prepolymer. Insensitive Munitions and Energetic Materials Symposium. Orlando, FL, USA (March 2003); (c) Paul, N.C., Desai, H., Cunliffe, A.V., Rodgers, M., Bull, H., and Leeming, W.B.H. (1995) An improved poly(GLYN) binder through end group modification. Proceedings of Joint International Symposium on Energetic Materials Technology, Phoenix, Arizona, USA, 52–60; (d) Paraskos, A.J., Dewey, M.A., and Edwards, W. (2010) One pot procedure for PolyGLYN end modification. US Patent 7,714,078 B2; (e) Leeming, W.B.H., Marshall, E.J., Bull, H., Rodgers, M.J., and Paul, N.C. (1995) An investigation into Poly (GLYN) cure stability. Proceedings of 27th Annual Conference of ICT, Karlsruhe, 99/1–99/5.

21 (a) Golding, P., Miller, R.W., Paul, N., and Richards, D.H. (1993) Preparation of di and polynitrates by ring opening nitration of oxetanes by dinitrogen pentoxide. *Tetrahedron*, **49**, 7051–7062; (b) Fischer, J.W. and Hollins, R.A. (1991) Synthesis of nitrato methyl methyl oxetane. US Statutory Invention Registration Publication No:H991. Published by United States Patent Office on 05-Nov-1991.

22 Manser, G.E. and Hajik, R.M. (1993) Method of synthesizing nitrato alkyl oxetanes. US Patent 5,214,166.

23 Liu, Y.L., Hsiue, G.H., and Chiu, Y.S. (1993) Studies on the polymerization mechanism of 3-nitratomethyl-3′-methyloxetane and 3-azidomethyl-3′-methyloxetane and the synthesis of their respective triblock copolymers with tetrahydrofuran. *J. Polym. Sci. A. Polym. Chem.*, **33**, 1607–1613.

24 (a) Pattison, D.B. (1957) Cyclic ethers made by pyrolysis of carbonate esters, *J.*

Am. Chem. Soc., **79**, 3455–3456;
(b) Vandenberg, E.J., Mullis, J.C., and Juvet, R.S. Jr (1989) Poly (3,3-bis (hydroxymethyl)oxetane) – an analog of cellulose: synthesis, characterization and properties. *J. Polym. Sci. A. Polym. Chem.*, **27**, 3083–3112.

25 Cumming, A. (1997) New directions in energetic materials. *J. Def. Sci.*, **1** (3) 319.

26 (a) Manser, G.E. (1982) Energetic copolymers and method of making same. US Patent 4,483,978; (b) Wardle, R.B. and Hinshaw, J.C. (1991) Cationic polymerization of cyclic ethers. US Patent 4,988,797; (c) Stewart, M.J. (1993) Process for the production of polyethers derived from oxetanes. US Patent 5,210,179; (d) Desai, H.J., Cunliffe, A.V., Hamid, J., Honey, P.J., Stewart, M.J., and Amass, A.J. (1996) Synthesis and characterization of α,ω-hydroxy and telechelic oligomers of NIMMO and GLYN. *Polymer*, **37** (15), 3461–3469; (e) Malik, A.A., Archibald, T.G., Carlson, R.P., and Manser, G.E. (1995) Polymerization of energetic cyclic ether monomers using $BF_3 \cdot THF$. US Patent 5,468,841.

27 (a) Farncomb, R.E. and Nauflett, G.W. (1997) Innovative polymer processing in carbon dioxide. *Waste Manage.*, **17**, 123–127; (b) Nauflett, G.W. and Farncomb, R.E. (2001) Nitration of organics in carbon dioxide. US Patent 6,177,033 B1.

28 Colclough, M.E. and Paul, N.C. (1996) Nitrated hydroxy terminated polybutadienes: synthesis and properties. *ACS Symp. Ser.*, **623**, 97–103.

29 Agrawal, J.P. (2005) Some new high energy materials and their formulations for specialized applications. *Propellants Explos. Pyrotech.*, **30** (5), 316–328.

30 Consaga, J.P. and Collignon, S.L. (1992) Energetic composites of cyclodextrin nitrate esters and nitrate ester plasticizers. US Patent 5,114,506.

31 (a) Ruebner, A., Statton, G.L., and Consaga, J.P. (2003) Polymeric cyclodextrin nitrate esters. US Patent 6,527,887 B1; (b) Ruebner, A., Statton, G.L., Robitelle, D., Meyers, C., and Kosowski, B. (2000) Cyclodextrin polymer nitrate. 31st International Annual Conference of ICT, 12/1–12/9; (c) Consaga, J.P. and Gill, R.C. (1998) Synthesis and use of cyclodextrin nitrate. 29th International Annual Conference of ICT, Karlsruhe, 5/1–5/5.

32 Phillips, R.W., Orlick, C.A., and Steinberger., R. (1955) The kinetics of the thermal decomposition of nitrocellulose. *J. Phys. Chem.*, **59**, 1034–1039.

33 (a) Jutier, J.J., Harrison, Y., Premont, S., and Prud'homme, R.E. (1987) A nonisothermal Fourier transform infrared degradation study of NC derived from wood and cotton. *J. Appl. Polym. Sci.*, **33**, 1359–1375; (b) Jutier, J.J. and Prud'homme, R.E. (1986) Thermal decomposition of nitrocelluloses derived from wood and cotton. A nonisothermal thermogravimetric analysis. *Thermochim. Acta*, **104**, 321–337.

34 Eisenreich, N. and Pfeil, A. (1983) Non-linear least squares fit of non-isothermal thermoanalytical curves. Reinvestigation of the kinetics of the autocatalytic decomposition of nitrated cellulose. *Thermochim. Acta*, **61**, 13–21.

35 (a) Chen, J.K. and Brill, T.B. (1991) Thermal decomposition of energetic materials. Part 51. Kinetics of weight loss from nitrate ester polymers at low heating rates. *Thermochim. Acta*, **181**, 71–77; (b) Chen, J.K. and Brill, T.B. (1991) Thermal decomposition of energetic materials 50. Kinetics and mechanism of nitrate ester polymers in high heating rates by smatch/FTIR spectroscopy. *Combust. Flame*, **85**, 479–488; (c) Brill, T.B. and Gongwer, P.E. (1997) Thermal decomposition of energetic materials 69. Analysis of the kinetics of nitrocellulose at 50 °C–500 °C. *Propellants Explos. Pyrotech.*, **22**, 38–44.

36 (a) Manelis, G.B., Rubtsov, Y.I., Smirnov, L.P., and Dubovitskii, F.I. (1962) Kinetics of thermal decomposition of pyroxylin. *Kinetica I Kataliz*, **3**, 42–48; Manelis, G.B., Nazin, G.M., Rubtsov, Y.I., and Strunin, V.A. (2003) *Thermal Decomposition and Combustion of Explosives and Propellants*, Taylor and Francis Group, p. 129.

37 Wolfrom, M.H., Frazer, J.H., Kuhn, L.P., Dickey, E.E., Olin, S.M., Hoffman,

D.O., Bower, R.S., Chaney, A., Carpenter, E., and McWain, P. (1955) The controlled thermal decomposition of cellulose nitrate. I. *J. Am. Chem. Soc.*, **77**, 6573–6580.

38 Bunyan, P.F. (1992) An investigation of the thermal decomposition of poly (3-nitratomethyl-3-methyloxetane). *Thermochim. Acta*, **207**, 147–159.

39 Kemp, T.J., Barton, Z.M., and Cunliffe, A.V. (1998) Mechanism of thermal degradation of pre-polymeric poly (NIMMO). *Polymer*, **40**, 65–93.

40 Bunyan, P.F. (1995) A study of the thermal decomposition of polyGLYN (poly(glycidyl nitrate)). Proceedings – International Symposium on Energetic Materials Technology, Phoenix, 147–152.

41 Bull, H., Bunyan, P.F., Cunliffe, A.V., Leeming, W.B.H., Marshall, E.J., and Rodgers, M.J. (1998) An investigation into the thermal stability of end-modified poly (Glyn). 29th International Conference of ICT, 89.1–89.12.

42 Stoltz, C.A. and Peiris, S.M. (2007) T-Jump/FTIR studies of poly-glycidyl nitrate (PGN) pyrolysis. AIP Conference on Shock Compression of Condensed Matter, 1263–1266.

43 Ling, P. and Wight, C.A. (1997) Laser photodissociation and thermal pyrolysis of energetic polymers. *J. Phys. Chem. B*, **101**, 2126–2131.

44 (a) Lengelle, G., Duterque, J., and Trubert, J.F. (2000) Physico-chemical mechanisms of solid propellant combustion, in *Solid Propellant Chemistry, Combustion and Motor Interior Ballistics*, Vol. 185 (eds V. Yang, T.B. Brill, and W.-Z. Ren), Progress in Astronautics and Aeronautics, pp. 287–334; (b) Duterque, J., Hommell, J., and Lengelle, G. (1985) Experimental study of double-base propellants combustion mechanisms. *Propellants Explos. Pyrotech.*, **10**, 18–25.

45 Yang, R., Thakre, P., Liau, Y.-C., and Yang, V. (2006) Formation of dark zone and its temperature plateau in solid-propellant flames: a review. *Combust. Flame*, **145**, 38–58.

46 (a) Roh, T.S., Tseng, I.S., and Yang, V. (1995) Effects of acoustic oscillations on flame dynamics of homogeneous propellants in rocket motors, *J. Propul. Power*, **11**, 640–650; (b) Tseng, I.S. and Yang, V. (1994) Combustion of a homogenous double base propellant in a rocket motor. *Combust. Flame*, **96**, 325–342.

47 Krause, H.H. (2005) New energetic materials, in *Energetic Materials Particle Processing and Characterization* (ed. U. Tiepel), Wiley-VCH Verlag GmbH.

48 Sutton, G.P. and Biblarz, O. (2001) *Rocket Propulsion Elements*, 7th edn, Wiley-VCH Verlag GmbH.

49 Kulkarni, A.R., Bhat, V.K., Phadke, S.P., and Nair, R.G.K. (1990) Simplified burn rate model for CMDB propellants. *Def. Sci. J.*, **40** (3), 255–262.

50 Kubota, N. and Masamato, T. (1976) Flame structures and burn rate characteristics of CMDB propellants. 16th International Symposium on Combustion. Pittsburgh, 1201–1209.

51 Bhalerao, M.M., Gautam, G.K., Subramanian, G.V., and Singh, S.N. (1996) Nitramine double base propellants. *Def. Sci. J.*, **46** (4), 207–214.

52 Asthana, S.N., Mundada, R.B., Phawade, P.A., and Shrotri, P.G. (1993) Combustion behaviour of advanced solid propellants. *Def. Sci. J.*, **43** (3), 269–273.

53 Yano, Y. and Kubota, N. (1986) Combustion of HMX-CMDB propellants (II). *Propellants Explos. Pyrotech.*, **11**, 1–5.

54 Beckstead, M.W., Puduppakkam, K., Thakre, P., and Yang, V. (2007) Modeling of combustion and ignition of solid propellant ingredients. *Prog. Energ. Combust. Sci.*, **33** 497–551.

55 Blomshield, F.S. (2006) Lesson learned in solid rocket combustion instability. Proceedings of AIAA Missile Sciences Conference Monterey, CA, USA.

56 Hau, Z., Feng, Z., Wang, E., and Han, P. (1992) The energy and pressure exponent of composite modified double-base propellants. *Propellants Explos. Pyrotech.*, **17** (2), 59–62.

57 Asthana, S.N., Ghavate, R.B., and Singh, H. (1990) Effects of high energy materials on the stability of CMDB propellants. *J. Hazard. Mater.*, **23**, 235–244.

58 Rice, D.D., Dubois, R.J., Lambert, R.S., and Thermal, A. (1968) Stability test for

composite modified double base propellants. *Explosivestoffe*, **11**, 245–249.
59. (a) Hartmann, K.O. and Morton, J.W. (1981) Alkoxy substituted aromatic stabilizers for cross-linked CMDB propellants. US Patent 4,299,636; (b) Berta, D.A. (1977) Composite modified double base propellant with metal oxide stabilizer. US Patent 3,905,846; (c) Asthana, S.N., Deshpande, B.Y., and Singh, H. (1989) Evaluation of various stabilizers for stability and increased life of CMDB propellants. *Propellants Explos. Pyrotech.*, **14**, 170–175; (d) Asthana, S.N., Divekar, C.N., and Khare, R.R., Shrotri, P.G. (1992) Thermal behavior of AP-based CMDB propellants with stabilizers. *Def. Sci. J.*, **42** (3), 201–204.
60. Baczuk, R.J. (1977) Molecular sieve containing stabilizing system for urethane cross-linked double base propellant. US Patent 4,045,261.
61. Divekar, C.N., Asthana, S.N., and Singh, H. (2001) Studies on combustion of metallized RDX-based composite modified double-base propellants. *J. Propul. Power*, **17** (1), 58–64.
62. Simpson, R.L., Urtiev, P.A., Ornellas, D.L., Moody, G.L., Scribner, K.J., and Hoffman, D.F. (1997) CL-20 performance exceeds that of HMX and its sensitivity is moderate. *Propellants Explos. Pyrotech.*, **22** (5), 249–255.
63. Nair, U.R., Gore, G.M., Sivabalan, R., Divekar, C.N., Asthana, S.N., and Singh, H. (2004) Studies on advanced CL-20 based double-base propellants. *J. Propul. Power*, **20** (5), 952–955.
64. Schoyer, H.F.R., Schnorhk, A.J., Mul, J.M., Gadiot, G.M.H.J.L., and Meulenbrugge, J.J. (1995) High performance propellants based on hydrazinium nitroformate. *J. Propul. Power*, **11**, 856–869.
65. Dendage, P.S., Sarwade, D.B., Asthana, S.N., and Singh, H. (2001) Hydrazinium nitroformate (HNF) and HNF based propellants: a review. *J. Energ. Mat.*, **19** (1), 14–78.
66. Dendage, P.S., Asthana, S.N., and Singh, H. (2003) Ecofriendly energetic oxidizer-hydrazinium nitroformate (HNF) and propellants based on HNF. *J. Ind. Chem. Soc.*, **80** (5), 563–568.
67. Lieb, R.J. and Heimerl, J.M. (1994) Characteristics of JAX gun propellants. Proceedings of the 6th International Gun Propellant and Propulsion Symposium. Parsippany.
68. (a) Leach, C., Debenham, D., Kelly, J., and Gillespie, K. (1998) Advances in poly NIMMO composite gun propellants. *Propellants Explos. Pyrotech.*, **23**, 313–316; (b) Debenham, D.F. (2001) Extrudable gun propellant composition. US Patent 6,228,190 B1.
69. Bohn, M.A. and Mueller, D. (2006) Insensitivity aspects of NC bonded and DNDA plasticizer containing gun propellants. Proceedings of Insensitive Munitions and Energetic Materials Technical Symposium (IMEMTS), Bristol, UK.
70. Atwood, A.I., Boggs, T.L., Curran, P.O., Parr, T.P., and Hanson Parr, D.M. (1999) Burning rate of solid propellant ingredients. *J. Propul. Power*, **15**, 740–747.
71. Schoyer, H.F.R., Korting, P.A.O.G., Veltmans, W.H.M., Louwers, J., v.d. Heijden, A.E.D.M., Keizers, H.L.J., and v.d. Berg, R.P. (2000) An overview of the development of HNF and HNF-based propellants. Proceedings of the 36th AIAA/ASME/SAE/ASEE Joint Propulsion Conference and Exhibit. Huntsville.
72. Cunliffe, A., Carina, E., Marshall, E., O'Day, E., and Wingborg, N. (February 2002) United Kingdom/Sweden Collaboration on ADN and PolyNIMMO/PolyGLYN formulation assessment. Report Number FOI-R-0420-SE. Issued by FOI-Swedish Defence Research Agency.
73. Weiser, V., Eisenreich, N., Baier, A., and Eckl, W. (2000) Burning behaviour of ADN formulations. *Propellants Explos. Pyrotech.*, **24** (3), 163–167.
74. Kubota, N. and Miyazaki, S. (1987) Temperature sensitivity of burning rate of ammonium perchlorate propellants. *Propellants Explos. Pyrotech.*, **12**, 183–187.
75. (a) Rao, K.P.C., Sikder, A.K., Kulkarni, M.A., Bhalerao, M.M., and Gandhe, B.R. (2004) Studies on BuNENA: synthesis

characterization and propellant evaluations. *Propellants Explos. Pyrotech.*, **29**, 93–98; (b) MSDS data sheet for double base propellant, Prepared by Western Powders Inc, Miles City, MT, USA. Revised on 15 November 2007. http://www.accuratepowder.com/data/AccurateDoubleBasePropellant.pdf; (c) Manning, T.G., Mishock, J., Adam, C., Kostka, J., Lieb, R., Leadore, M., Worrell, D.A., Hollins, R., and Ritchie, S.J. (2007) Environment friendly propellant for large caliber training rounds. Proceedings of Insensitive Munitions and Energetic Materials Symposium, Miami, USA.

76 (a) Jet Propulsion Laboratory (1957) High-Performance Polyglycidyl Nitrate-Polyurethane Propellants, 33 pages, Mar. 29; (b) Defense Technical Information Center (DTIC), (1957) Document No. AD 144756; (c) Defense Technical Information Center (DTIC), (1957) Document No. AD 139462.

77 Willer, R.L. and McGrath, D.K. (1998) High performance large launch vehicle solid propellants. US Patent 5,801,325.

78 Menke, K., Bohn, M., and Kempa, P.B. (2006) AN/Poly GLYN propellant formulations–approaches to less sensitive rocket propellants. Proceedings of Insensitive Munitions and Energetic Materials Symposium, Bristol, UK.

79 Menke, K., Gunser, G., Schnoering, M., Mauβ, J.B., and Weiser, V. (2003) Burning rate enhancement of AN-propellants by TAGN and GZT. Proceedings of 34th International Annual Conference of ICT, Karlsruhe, 22/1.

80 Benziger, T.M. (1973) High energy plastic bonded explosive. US Patent 3,778,319.

81 (a) Persson, P.A., Holmberg, R., and Lee, J. (1994) *Rock Blasting and Explosives Engineering*, CRC Press; (b) Humphrey, J.R. (May 1996) Safety handling characteristics of LX-04-01, UCRL-ID-124086; Published by Technical Information Department, Lawrence Livermore National Laboratory (LLNL), University of California, USA;

(c) Simpson, L.R. and Frances Foltz, M. (January 1995) UCRL-ID-119665, Published by Technical Information Department, Lawrence Livermore National Laboratory (LLNL), University of California, USA.

82 (a) Hollands, R., Fung, V., and Burrows, K.S. (2004) High performance polymer bonded explosive containing PolyNIMMO for metal accelerating applications. Proceedings of NDIA Insensitive Munitions & Energetic Materials Symposium, San Francisco, USA; (b) Hatch, R. and Braithwaite, P. (2007) IM and qualification testing of cast cure CL-20 explosive DLE-C038. Proceedings of Insensitive Munitions and Energetic Materials Symposium. Miami, USA.

83 Smith, M.W. and Cliff, M.D. (1999) NTO based explosive formulations: a technology review. Technical Report, 0796, DSTO.

84 Isler, J. (1994) The class/division 1.6: an analysis of what are EIDS high explosives and which explosive effects 1.6 articles are likely to produce. Proceedings of the 26th Explosive Safety Seminar, Miami, USA.

85 (a) Karlsson, S., Ostmark, H., Eldsater, C., Carlsson, T., Bergman, H., Wallin, S., and Pettersson, A. (2002) Detonation and sensitivity properties of FOX-7 and formulations containing FOX-7. Proceedings of the 12th International Detonation Symposium. Sandiego, USA; (b) Eldsater, C., Pettersson, A., and Wanhatalo, M. (2004) Formulation and testing of a Comp B replacement based on FOX-7. Proceedings of Insensitive Munitions & Energetic Materials Symposium. San Francisco, USA.

86 Busby, A.J. (2006) Pressable PBX formulation based on FOX-7. Proceedings of Insensitive Munitions and Energetic Materials Symposium, Bristol, UK.

87 Smith, L.C. (1977) On the problem of evaluating safety of an explosive. Proceedings of the Conference on the Standardisation of Safety and Performance Tests. Dover, NJ, USA.

4
Energetic Thermoplastic Elastomers

4.1
Introduction

Thermoplastic elastomers (TPE) are biphasic materials possessing the combined properties of glassy or semi-crystalline plastics and soft elastomers. This property enables rubbers to be processed as thermoplastics [1]. They are single macromolecules in which discrete thermoplastic segments, capable of forming rigid nanoscale domains or channels, are covalently bonded to rubbery segments that provide the soft matrix in which the rigid domains reside [2]. Below the glass transition temperature (T_g) of the rigid segments, the rigid domains form three-dimensional networks of physical cross-link sites, which act as cross-links for the soft matrix (Figure 4.1). Hence TPEs exhibit elastomeric properties if the volume of low T_g soft blocks is large [3]. The hard domains serve as thermally reversible cross-links, which help to process the TPE as a thermoplastic through high-throughput thermoplastic processes such as melt extrusion and injection molding. The most well known elastomers of this type are the polystyrene (PS)-block-polybutadiene (PB)-block-PS tri-block copolymers, sold commercially as Kraton.

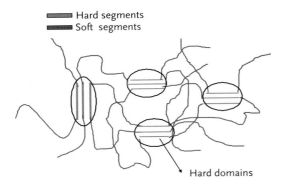

Figure 4.1 Schematic of the microphase separated structure of thermoplastic elastomer.

Energetic Polymers: Binders and Plasticizers for Enhancing Performance, First Edition.
How Ghee Ang and Sreekumar Pisharath.
© 2012 WILEY-VCH Verlag GmbH & Co. KGaA. Weinheim. Published 2012 by WILEY-VCH Verlag GmbH & Co. KGaA

TPEs based on segmented polyurethanes (Estane 5703) and block copolymers of styrene and ethylene/butylene (Kraton G-6500) are widely used as binders in propellant and explosive formulations [4]. There are significant advantages to using thermoplastic elastomeric binders over conventional cast cure binders, such as HTPB (hydroxy terminated polybutadiene). A thermoplastic binder can be easily and safely processed compared with the conventional binders, and the elastomeric properties of the binder permit it to absorb a part of the impact energy, thus reducing the overall shock sensitivity. Furthermore, unlike the situation with chemically cross-linked conventional binders, TPE formulations are recyclable by simply melting the material [5].

Energetic thermoplastic elastomers (ETPE) are prepared by copolymerizing energetic polymers. ETPEs are expected to bestow the high performance and insensitivity attributes of energetic polymers to formulations in addition to the processing and recycling advantages of TPEs mentioned earlier. They have emerged as potential energetic binder candidates and have been successfully used in formulations of new (LOVA) propellants [6].

An ideal ETPE binder is an ABA block copolymer, where A blocks have suitable melting temperatures and B blocks have a lower glass transition temperature. ETPEs that have been developed for binder applications include the copolymers BAMO/AMMO, BAMO/GAP, BAMO/NIMMO, and segmented copolyurethane elastomers based on GAP or Poly(NIMMO).

In BAMO-based ETPEs, the energetic oxetane polymer BAMO, which melts at 83 °C, constitutes the hard segment and AMMO, GAP, and NIMMO, having lower glass transition temperatures, make up the soft segment. All these polymers are expected to melt in the temperature range between 80 and 100 °C, which is ideal for melt casting of propellant and explosive formulations. In the GAP-based Segmented copolyurethane ETPE, the hard segment is generated by the formation of hydrogen bonds between the first urethane group of one linear copolymer chain with the second urethane group of another copolyurethane chain [6].

In general, the properties of the ETPE for binder applications depend on the molecular weights of the individual blocks and also on the ratio of the blocks to each other. The molecular weight of the hard segment ranges from 3000 to 8000 g/mol and that of the soft segment is in the range of 3000–15 000 g/mol. The weight ratio of the hard blocks to soft blocks ranges from 15 : 85 to about 40 : 60 [6, 7].

This chapter discusses the preparation and properties of important ETPEs used for binder applications and also some of the promising formulations using ETPE binders.

4.2
Preparation of Energetic Thermoplastic Elastomers

Broadly, two routes have been adopted for the synthesis of ABA ETPE systems. These are:

1. Coupling of telechelic low T_g/amorphous prepolymer with monofunctional high T_g/crystalline prepolymer as terminal hard blocks by adopting a block-linking approach.
2. Polymerization of monomers through sequential addition by utilizing a living polymerization approach.

In the block-linking approach, the middle B block is functionalized at both of its ends by a linking moiety (e.g., isocyanate), which subsequently reacts with A blocks to form the ABA block copolymer [7]. The reaction is represented as:

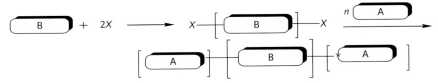

In the sequential addition approach, the monomer A undergoes cationic polymerization to form an A block with an active end. The second monomer B is added and the polymerization of B monomer is initiated by the active end of the A block. This results in the formation of an AB diblock polymer with an active end. On adding extra A monomer, the ABA tri-block copolymer is formed [7]. Generally the reaction is represented as:

Alternatively, a di-functional initiator is used to initiate the polymerization of the middle B block. Upon addition of monomer A, the polymerization would proceed from the both the active ends of the B block [7]. The reaction could be represented as:

Among the aforementioned methods the block-linking approach has been widely exploited to synthesize high molecular weight ETPEs [8].

In a representative example of the application of this approach, Xu et al. [9] synthesized di-and tri-block copolymers of BAMO and NIMMO. The linking of the blocks was carried out by toluene 2,4-diisocyanate (TDI) using dibutyltin dilaurate (DBTDL) as catalyst as shown in Scheme 4.1.

The average molecular weight of the synthesized tri-block copolymer was 27 000 g/mol. It was observed that the thermal and spectroscopic properties of the di- and tri-block copolymers were similar. Both polymers melt at 82 °C, and show a T_g of −3 °C as shown in Figure 4.2.

The synthetic procedure of linking the telechelic energetic polymers by isocyanate linkages was further improvised by Sanderson and coworkers [9, 10] by

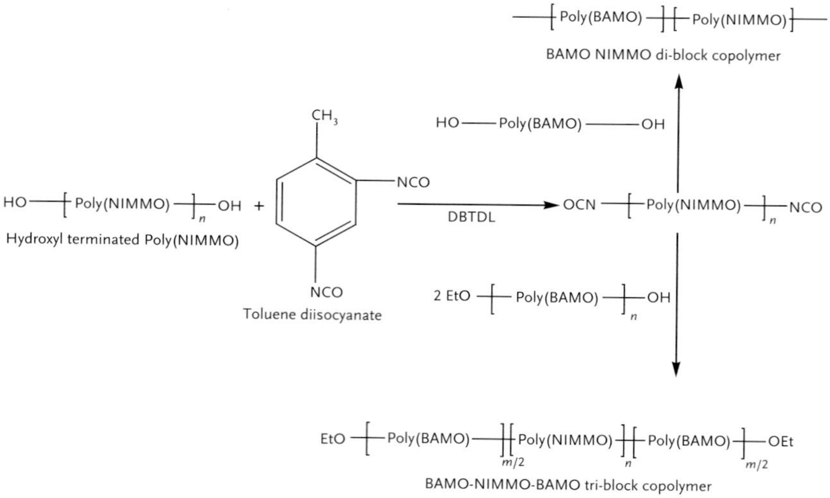

Scheme 4.1 Synthesis of BAMO-NIMMO ETPE using block-linking approach [9].

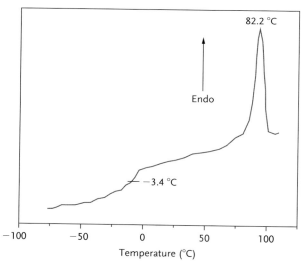

Figure 4.2 DSC curve of BAMO-NIMMO-BAMO copolymer. Figure reproduced with permission of Wiley-VCH from Ref. [9].

including oligomeric urethane linkages for connecting two blocks, which helps to enhance the softening temperature of the ETPE (Scheme 4.2). The urethane linkage is formed by the reaction of the end-capped isocyanate blocks and bifunctional diols, such as ethylene glycol, propylene glycol, and cycloaliphatic diols, for example, 1,4-cyclohexane dimethanol.

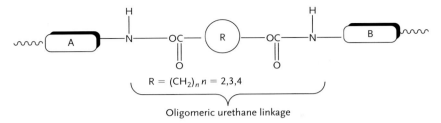

Scheme 4.2 Oligomeric urethane linkage connecting two copolymer blocks.

Table 4.1 Typical molecular weights of representative ETPEs prepared by the block-linking technique.

ETPE	BAMO-GAP	BAMO-NIMMO	BAMO-AMMO
M_w	219500	105900	15500
M_n	28440	15020	8100
Dispersity	7.7	7.1	1.92

Typical molecular weights of various ETPEs prepared by the block-linking approach are presented in Table 4.1 [9, 11].

From this table, it is clear that ETPEs with higher molecular weights are possible with the block-linking technique. However, the ETPE polymers prepared have high polydispersity.

Sequential addition of monomers is also pursued as an alternative approach to block linking for ETPE preparation. This approach eliminates the use of toxic isocyanates in the preparation of ETPE binders. Also, the sequential addition approach is expected to provide ETPE with uniform macromolecular architecture and therefore lower polydispersity.

Kimura et al. [12] successfully used this approach to synthesize the copolymers of BAMO and NIMMO using a 1,4-butanediol/boron trifluoride etherate initiating system. The copolymerization of BAMO with NIMMO occurs as an ideal system, with the copolymer having the exact composition as the proportions in the monomer feed. The microstructures of the synthesized copolymer were observed to be randomly arranged. The molecular weights were in the range of 2000–3000 g/mol depending on the mole ratio of BAMO to NIMMO.

Pisharath et al. [13] prepared the BAMO-GAP-BAMO ETPE by using cationic ring opening polymerization as shown in Scheme 4.3. Synthesis of PBCMO-PECH-PBCMO copolymer, the halogen precursor to BAMO-GAP-BAMO ETPE, accomplished by the ring opening polymerization of 3,3-bis(chloromethyl) oxetane (BCMO) in the presence of PECH diol as a macro-initiator and boron trifluoride/dimethyl ether complex catalyst.

The PECH-PBCMO copolymer was converted into GAP-PBAMO copolymer by an azidation reaction using sodium azide. BAMO-GAP ETPE with a molecular

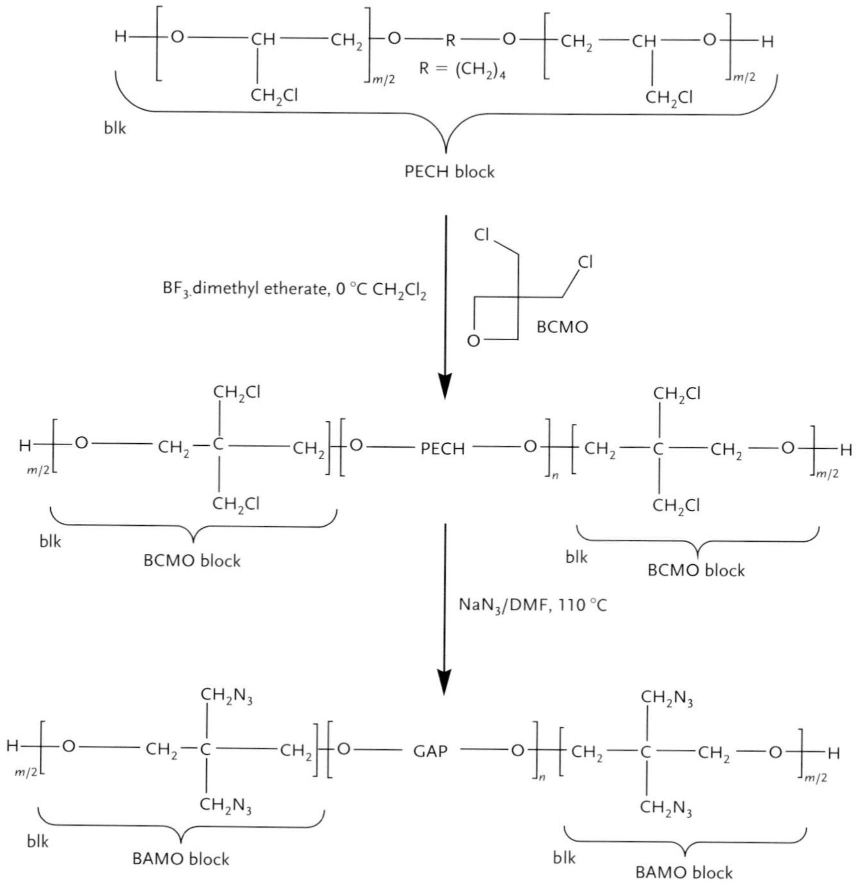

Scheme 4.3 Synthetic scheme for BAMO-GAP-BAMO ETPE.

weight of 6000 g/mol and polydispersity of 1.4 was synthesized by this route [13]. The copolymer consists of 72% GAP segments and 28% Poly(BAMO) segments. The differential scanning calorimetry (DSC) curve of the ETPE in the range from −50 to 100 °C shows two glass transition temperatures at −35 °C (due to the GAP block) and −16 °C (due to the PBAMO block), and softening at 66 °C (Figure 4.3).

Morphological observation by hot stage polarizing optical microscopy shows fine crystallites of the BAMO hard segment dispersed uniformly in the amorphous GAP homopolymer (Figure 4.4).

Kawamoto *et al.* [14] prepared random copolymers of GAP and Poly(BAMO) by copolymerization of epichlorohydrin and 3,3-bis(bromomethyl) oxetane (BBrMO) using 1,4-butanediol as initiator and boron trifluoride etherate as catalyst, followed by azidation with sodium azide in dimethyl sulfoxide (DMSO) solvent (Scheme 4.4).

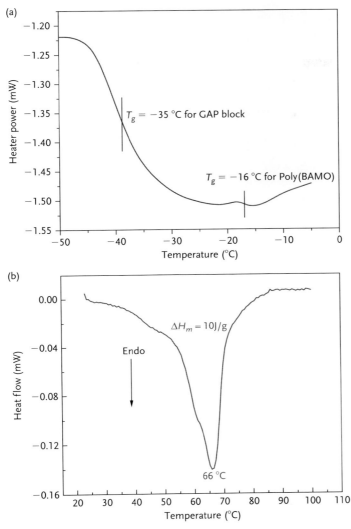

Figure 4.3 DSC curves of BAMO-GAP-BAMO ETPE showing (a) glass transition and (b) softening temperatures (unpublished results H.G. Ang and S. Pisharath).

The objective of the work was to prepare a fully amorphous polymer having higher energy than the GAP homopolymer. The synthesized copolymer contained both cyclics and linear chains and the M_n of the linear chains ranged from 1380 to 1460 g/mol depending on the ratio between GAP and Poly(BAMO). All the copolymers have higher decomposition energy as compared with the GAP homopolymer.

Talukder et al. [15] used a difunctional initiator, p-bis(α,α-dimethylchloromethyl) benzene (p-DCC) to synthesize BAMO-NIMMO-BAMO tri-block copolymer. In the first step, p-DCC was reacted with silver hexafluoroantimonate (AgSbF$_6$) to

Figure 4.4 Polarizing optical micrograph of GAP-Poly(BAMO) ETPE (unpublished results H.G. Ang and S. Pisharath).

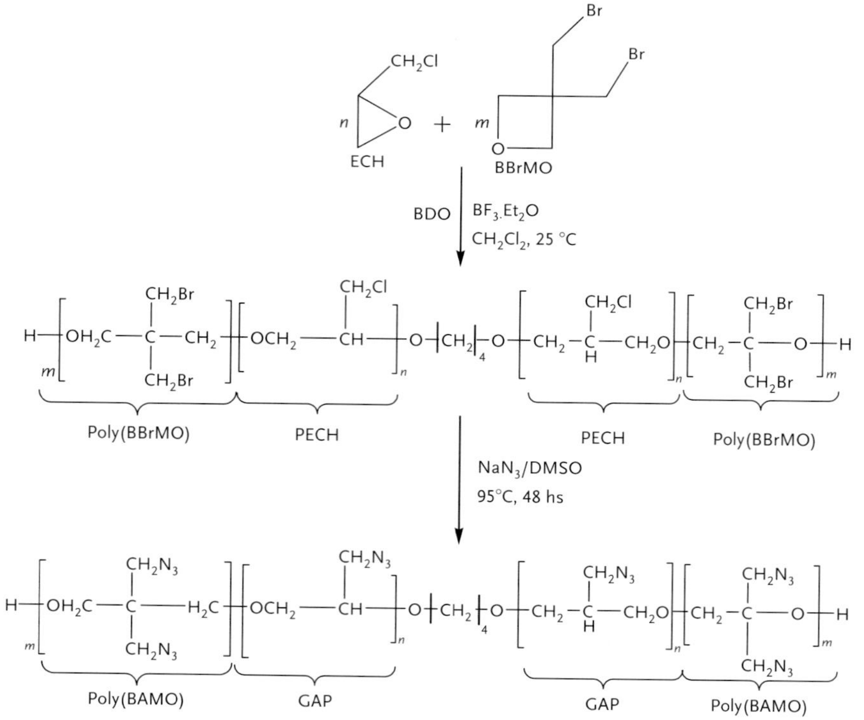

Scheme 4.4 Synthetic scheme of preparation of random copolymers of GAP and Poly(BAMO) [14].

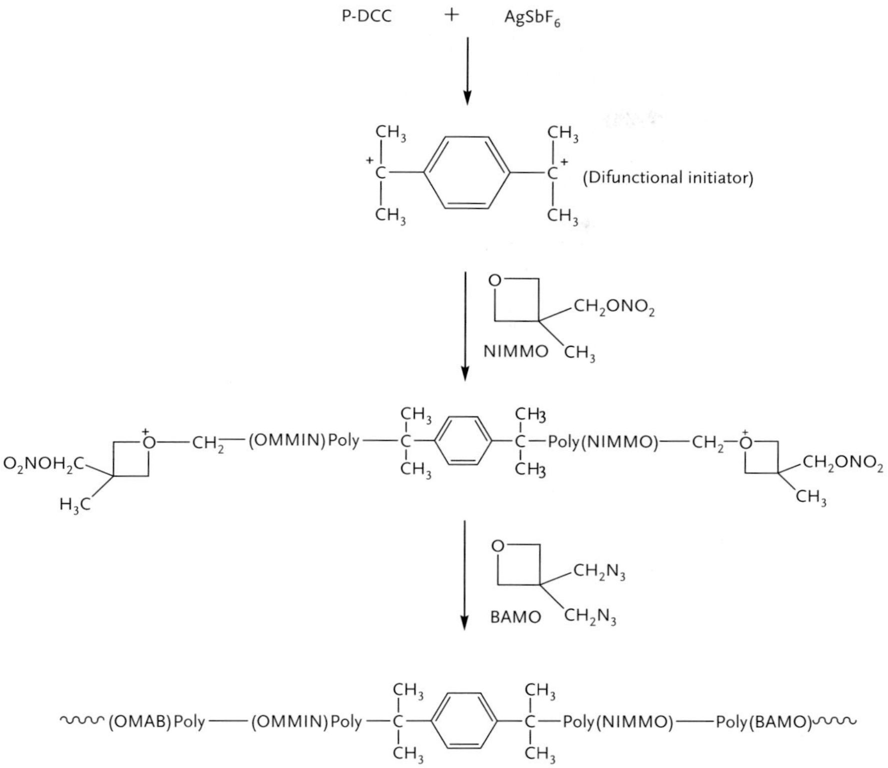

Scheme 4.5 Synthesis of BAMO-NIMMO-BAMO ETPE using difunctional initiator p-DCC.

generate a carbonium ion. Polymerization of NIMMO is initiated by this carbonium ion, which generates a Poly(NIMMO) molecule with a carbo-oxonium ion at its end. Polymerization of BAMO is initiated by this carbo-oxonium ion, which grows on either sides of the Poly(NIMMO) molecule (Scheme 4.5).

The reaction was carried out at −70 °C and a copolymer with an M_n of 223000 kg/mol was obtained with a dispersity index of 1.2. The copolymer also exhibited a glass transition temperature of ∼−30 °C and melting-point of 57 °C, as shown in the DSC thermogram (Figure 4.5).

Hsiue et al. [16] utilized another difunctional initiator; triflic anhydride for the synthesis of tri-block polymers of NIMMO with tetrahydrofuran (THF). Poly(THF) forms the crystalline middle block, with the Poly(NIMMO) segments forming the amorphous end blocks. A Poly(THF) chain with two living propagating chain ends was first obtained from the polymerization of THF using triflic anhydride difunctional initiator at 10 °C. After removing the excess of THF, NIMMO was added to the reaction mixture at 0 °C to continue the chain propagating reaction. Copolymers with weight average molecular weights (M_w) ranging from 12000 to 40000 kg/mol with an average polydispersity index of 1.2 were obtained. The DSC thermogram of the

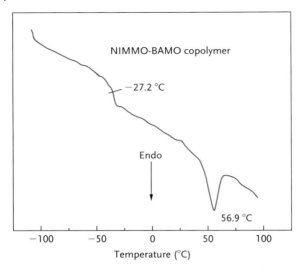

Figure 4.5 DSC profile of BAMO-NIMMO-BAMO copolymer. Reproduced with permission of Wiley from Ref. [15].

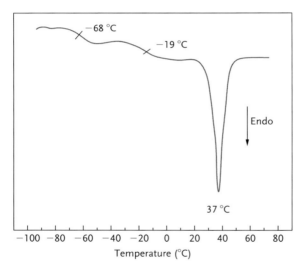

Figure 4.6 DSC profile of NIMMO-THF-NIMMO ETPE. Reproduced with permission of Wiley from Ref. [16].

copolymer shown in Figure 4.6 exhibits two T_g values (-68 and -19 °C) and a melting peak at 37 °C.

In a similar way, AMMO-THF-AMMO copolymers were also prepared [16]. The M_w of the prepared copolymers range from 3700 to 33000 kg/mol with an average polydispersity index of 1.2. The AMMO-THF-AMMO copolymer shows a melting peak at 43.1 °C. The melting-points of both AMMO-THF and NIMMO-THF copolymers are low and are not suitable for melt-cast application.

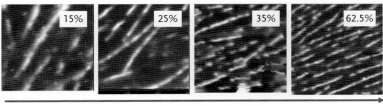

Figure 4.7 Chemical structure of GAP macro-initiator [17].

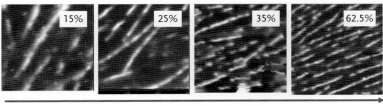

Figure 4.8 AFM images of annealed BAMO-AMMO copolymers with increasing %BAMO content [18]. Reprinted with permission of Dr. Pamela Kaste, US Army Research Laboratory.

Al-Kaabi and Van Reenen controlled radical polymerization (CRP) to prepare well defined ETPEs containing GAP and poly(methyl methacrylate) (PMMA) segments [17].

The PMMA serves as the thermoplastic segment with high oxygen content. In the preparative route, firstly a macro-initiator was prepared by attaching a dithiocarbamate initiator with a GAP segment (Figure 4.7). The macro-initiator was then used to photo polymerize methyl methacrylate (MMA) to form a PMMA-g-GAP copolymer having low polydispersity. The synthesized binder exhibited the decomposition behavior characteristics of GAP and was found to be compatible with RDX (Research Department Explosive) in vacuum thermal stability tests.

ETPEs are usually characterized by a variety of experimental techniques, such as nuclear magnetic resonance (NMR), gel permeation chromatography (GPC), differential scanning calorimetry (DSC), and atomic force microscopy (AFM) [18]. AFM is a unique tool to visualize the lamellae and spherulites formation induced by the crystalline phase, which controls the modulus of ETPE propellants.

As shown in Figure 4.8, the structure of individual crystalline lamella of BAMO units of the copolymer could be clearly visualized as light elongated structures with high aspect ratios. With increasing BAMO content in the copolymer, the concentration of lamellae increases and the feature becomes thinner [18].

4.3
Thermal Decomposition and Combustion of ETPEs

Only a limited number of research reports have been published on the thermal decomposition behavior of ETPEs. Kimura and Oyumi [19] compared thermal decomposition of BAMO-NIMMO ETPE with that of BAMO-AMMO. It was observed that for BAMO-NIMMO, thermal decomposition of the BAMO unit is accelerated by the decomposition of the NIMMO unit. On the other hand, for the BAMO-AMMO copolymer, the AMMO unit does not affect the thermal decomposition of BAMO.

Liu *et al.* [20] studied the thermal characteristics of copolymers of tetrahydrofuran (THF) with BAMO, AMMO, and NIMMO. It was observed that the decomposition enthalpies were dependent on the energetic group contents of the polymers and independent of whether the copolymer is of a block or random type. Copolymers containing NIMMO groups show the largest decomposition enthalpy, implying that the nitrato groups are more energetic than the azido groups. However, copolymers based on NIMMO, in spite of having a higher decomposition enthalpy, decompose at lower temperatures as compared with copolymers containing BAMO or AMMO. It is well known that the BAMO and AMMO based polymers show two-stage weight loss profiles in TGA experiments due to distinct weight losses from the energetic groups and the polymer backbone. Interestingly, the copolymers of NIMMO do not show the two-stage weight loss profiles, presumably due to the high decomposition enthalpy of the NIMMO, resulting in the simultaneous decomposition of the nitrato groups and the polymer backbone.

In order to study the effect of ageing on thermal decomposition, Chang *et al.* [21] compared the thermal decomposition behavior of the aged THF copolymers of AMMO or NIMMO with that of aged Poly(THF). The polymers were aged at 25 °C for 400 days. The activation energy of degradation was found to be lower for the aged copolymers as compared with that for Poly(THF) without ageing. The lower activation energy of the copolymers is ascribed to the presence of aged Poly(THF) segments containing weak bonds consisting of hydroperoxide groups, which undergo homolysis to hydroxyl and alkoxy radicals. The rearrangement of the radicals leads to chain scission and degradation.

Pisharath and Ang [22] studied the thermal decomposition kinetics of the GAP-Poly(BAMO) ETPE and applied model-free kinetics to obtain the activation energy of thermal decomposition.

The DSC curve of the ETPE shown in Figure 4.9 exhibits exothermic decomposition maximum at 229 °C with a decomposition enthalpy of 1400 J/g followed by a broad shoulder peak at 298 °C accounting for a decomposition enthalpy of 50 J/g. This behavior is different from the decomposition of either GAP or Poly(BAMO) homopolymers, which show single decomposition exotherms (see Figure 4.11, Chapter 2).

The FTIR spectra of the DSC residues collected at 250 °C confirm that one part of the azido groups in the copolymer decomposes at a higher temperature (Figure 4.10). The degraded residue of GAP-Poly(BAMO) ETPE shows the characteristic bands of azido groups at 2100 and 1270 cm^{-1} with reduced intensity. These bands

4.3 Thermal Decomposition and Combustion of ETPEs

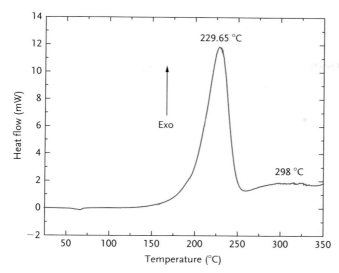

Figure 4.9 DSC profile of the GAP-Poly(BAMO) ETPE [22]. Reprinted with permission from Elsevier.

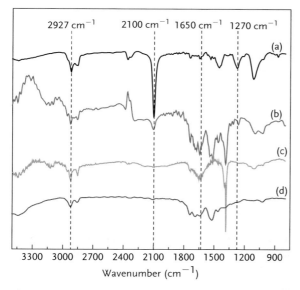

Figure 4.10 Comparison of FTIR spectra of (a) GAP-Poly(BAMO) copolymer, with thermally degraded residues of (b) GAP-Poly(BAMO) copolymer, (c) Poly(BAMO), and (d) GAP. The degraded residues were collected at 250 °C [22]. Reprinted with permission from Elsevier.

are completely absent for the degraded Poly(BAMO) and GAP homopolymers. However, a new peak at 1650 cm^{-1} appears in the FTIR spectra of the degraded polymer residues signifying the presence of imine intermediates.

FTIR analysis confirms that the residual azido groups in the copolymer after the first-stage decomposition are responsible for the second-stage decomposition observed in the GAP-Poly(BAMO) copolymer.

Advanced isoconversional analysis proposed by Vyazovkin and Dollimore [23] was utilized for calculating a model-free estimate of activation energy (E) as a function of the degree of decomposition (α). The results of the isoconversional analysis for GAP homopolymer and GAP-Poly(BAMO) are compared in Figure 4.11. The double decomposition behavior observed for ETPE in the DSC results could be identified as two regions of different activation energy in the dependence plot. The first region between conversions $\alpha = 0.03$ and $\alpha = 0.36$ with E_α of ~145 kJ/mol and the second region with E_α of ~220 kJ/mol to ~270 kJ/mol between $\alpha = 0.62$ and $\alpha = 0.9$.

On the other hand, GAP polymer decomposes with almost a constant activation energy, which is largely independent of conversion. Thus the azido groups in the GAP-Poly(BAMO) ETPE decompose in two stages at different activation energies. Previous studies on thermal characteristics of azido copolymers using model-fitting methods have suggested that the activation energy of azido decomposition is unaffected by copolymerization [24]. A more rigorous model-free approach indicates that activation energy of decomposition of the azido groups is affected by the copolymerization process.

The shape of the activation energy dependence curve throws light on the mechanism involved during the thermal decomposition process. In the literature, for degradation of linear polymers in the absence of oxygen, the increasing function of E_α dependence as observed for ETPE denotes a competition between

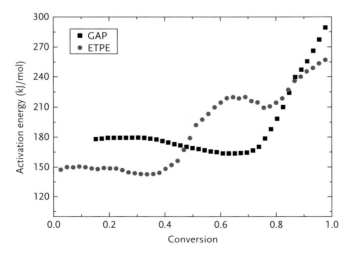

Figure 4.11 Comparison of activation energy dependence profiles of GAP-Poly(BAMO) ETPE with GAP calculated by Vyazovkin's isoconversional analysis [22]. Values for the plot taken with permission of Elsevier.

the decomposition of individual macromolecules and intermolecular associates formed during the course of the reaction [25, 26]. In the decomposition of GAP-Poly(BAMO) ETPE, the possible competition is between the decomposition of linear copolymer chains and the intermolecular associates formed from the imine intermediates usually formed during the decomposition of azido polymers [27, 28].

Sell *et al.* [29] used TGA at heating rates of between 0.5 and 10 °C/min to study the decomposition kinetics of the AP propellant of BAMO-AMMO ETPE. The propellant showed multi-step degradation kinetics, and isoconversional analysis was performed to observe any dependency of the activation energy on the extent of decomposition.

Figure 4.12 compares the activation energy dependencies of pure AP with that of its propellants with BAMO-AMMO and HTPB binders. Unlike HTPB/AP propellants, which show increasing dependencies of activation energy with respect to conversion [29], the BAMO-AMMO propellant has fairly constant activation energy (120 kJ/mol) for its major decomposition step. The constancy of effective activation energy suggests that the overall kinetics of this step is limited by a single reaction. The activation energy dependence of pure AP matches with that of BAMO-AMMO/AP propellant, indicating that thermal decomposition of the propellant is controlled by the decomposition of AP. The activation energy dependence of the HTPB propellant lies higher than that for pure AP, indicating that the kinetics of this step are determined by the decomposition of HTPB. Thus variations of the binder in the formulation led to the operation of different mechanisms in the propellant thermal decomposition kinetics.

Lee and Litzinger [31] have extensively studied the thermal decomposition of BAMO-AMMO ETPE in the presence and absence of titanium dioxide (TiO_2) with the heat fluxes delivered from a CO_2 laser, which simulated combustion

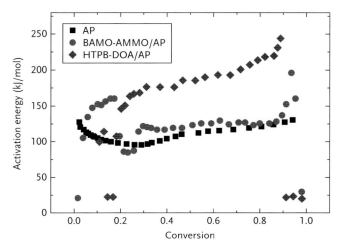

Figure 4.12 Comparison of activation energy dependencies for AP and its propellants with HTPB and BAMO-AMMO polymer binders. Values for the plot are taken with permission of Elsevier and American Chemical Society (Copyright 1999) from Refs. [29] and [30].

conditions. The decomposition studies were focused on the measurements of surface temperature and the nature of gaseous species evolving out of the process and arriving at the mechanism of thermal decomposition. ETPE in the absence of TiO_2 was found to undergo simultaneous decomposition of the backbone structure, as evidenced by the large amounts of formaldehyde, water, and carbon monoxide evolved along with the nitrogen and hydrogen cyanide from the side-chain decomposition. The relative concentrations of the evolved gas species were constant in the gas phase, indicating that the reactions are confined to the condensed phase. The lack of gas-phase reaction was also indicated by a relatively constant temperature profile value ranging from 800 to 950 K. With the addition of TiO_2 to ETPE, the reactions are still restricted to the condensed phase, but an increase in the mole fraction of ammonia was observed in the decomposition gases. The above observations indicate that the combustion of the azido polymer based ETPE is driven by reactions in the condensed phase, similar to the case with azido homopolymers.

Lee *et al.* [32] also investigated the effect of AMMO segments on the combustion of BAMO-AMMO copolymer. The existence of AMMO segments in the copolymer was found to dilute the surface and gas temperatures of pure BAMO by 200 K due to the lower energy content of AMMO. Previous literature on thermal decomposition of copolymers of BAMO [24] reported that the AMMO segment in the BAMO-AMMO copolymer does not affect the thermal decomposition of BAMO as such, which could be related to the low heat rates at which the experiments were conducted.

Braithwhite and coworkers [33] measured the burn rates of different ETPEs containing 25% BAMO as the hard segment using a strand burner (Figure 4.13).

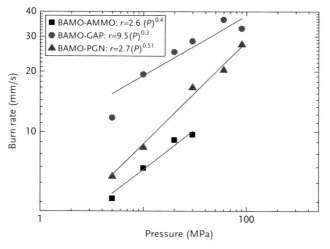

Figure 4.13 Comparison of burn-rate profiles of different ETPEs. Values for the plot are taken from Ref. [33].

The azido-based ETPEs have been found to have higher burning rates than the nitropolymer-based ETPEs. Furthermore, as AMMO has much less energy than any of the other energetic polymers, the ETPEs that contain AMMO have lower burning rates. Because of the large and wide range of burn rates available, these polymers are used as formulators of propellants for the purpose of tailoring the burn rates. It is noteworthy that the burn-rate exponents of all the ETPEs are good for propellant applications.

4.4
Combustion of ETPE Propellant Formulations

Combustion characteristics of propellant compositions of ETPEs have been studied in the literature. Kimura and Oyumi [34] studied the combustion characteristics of composite propellants of BAMO-NIMMO with HMX, AP, and AN.

Various combinations of catalysts based on lead, carbon black, copper chromite, and iron were attempted for augmentation of burn rates. Among the propellant compositions, the AP-based propellants showed plateau burning characteristics that were faster than for HMX- and AN-based propellants (Figure 4.14). The AP propellant has a specific impulse potential of 265 s. The combined catalysts of lead compounds and carbon black, or copper chromite and organic iron compound (FeB), were found to be effective in burn-rate augmentation and pressure-exponent modification.

Russian research groups have studied the burn rates, temperature profiles, burn-surface temperatures, and chemical constitution and thermal structures of various pseudo-propellants of the BAMO-AMMO copolymer with various

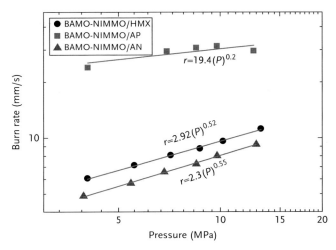

Figure 4.14 Burn-rate profiles of catalyzed propellants of BAMO-NMMO ETPE with oxidizers; HMX, AP, and AN. Values for the plot are taken with permission of Taylor and Francis Inc. from Ref. [34].

nitramines, including that with CL-20 [35, 36]. The investigations were conducted using a combination of microcouple techniques and molecular beam mass spectrometry at different pressures and temperature ranges.

Similar to the situation with GAP nitramine propellants, the burning rate of BAMO-AMMO/nitramine is controlled mainly by the heat release in the condensed phase just under the burning surface or immediately on the burning surface [35].

The nature of the binder influences the burn rate, flame temperature, and flame temperature profiles of the propellants. As shown in Figure 4.15, replacement of HTPB binder in HMX based pseudo-propellant with energetic GAP or BAMO-AMMO binder has a clear advantage of improving the burn rate of formulations (Figure 4.15). The GAP and BAMO-AMMO formulations appear to have similar burn-rate profiles. However, the power law fit equations indicate that incorporation of BAMO-AMMO copolymer in place of GAP decreases the burn rate of the formulation with an increase in the magnitude of the pressure-rate exponent [36].

At a given pressure, the final flame temperatures of GAP/HMX, BAMO-AMMO/HMX, and HTPB/HMX pseudo-propellants are 2580, 2475, and 2300 K, respectively [35, 36]. The higher flame temperatures of GAP and BAMO-AMMO formulations predict higher specific impulses for energetic binder based propellant formulations.

As illustrated in Figure 4.16, the final flame temperatures are reached at shorter distances from the burning surface for the energetic binder propellants. These are 0.5, 0.9, and ~1.1 mm for the GAP, BAMO-AMMO, and HTPB pseudo-propellants, respectively. This feature should help in the enhanced heat feedback from the gas phase to the burning surface. Both GAP and BAMO-AMMO based pseudo-propellants have narrower flame zones as compared with that of the HTPB pseudo-propellant [36].

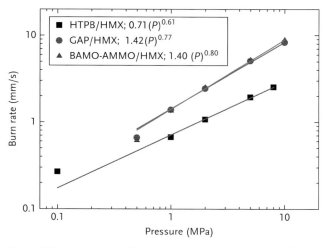

Figure 4.15 Comparison of burn-rate profiles of pseudo-propellants of HMX with binders GAP, BAMO-AMMO, and HTPB. Values for the plot are taken from Refs. [35(b)] and [35(c)].

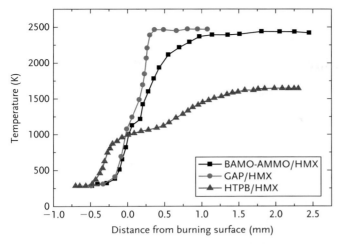

Figure 4.16 Flame temperature profiles of pseudo-propellants of HMX with GAP, BAMO/AMMO and HTPB at 1 MPa. Values for the plot are taken with permission of Elsevier from Refs. [35(c)] and [36].

The pseudo-propellants of HMX with GAP and BAMO-AMMO differ in the chemical composition of the burning surface and flame zones, as estimated by molecular beam mass spectrometry [36]. A higher concentration of N_2 was observed at the burning surface of BAMO-AMMO/nitramine pseudo-propellants due to the higher content of azido groups in the copolymer. Carbon residues were observed on the burning surface. In contrast to the GAP/HMX flame, it is believed that the experimental data on the thermal profile and chemical composition of the flame zones will provide essential background for the development of robust combustion models for ETPE based propellant formulations.

4.5
Performance of ETPE Based Propellant Formulations

ETPE binders are being widely used in experimental formulations for advanced gun propellants for extended range munitions and advanced gun systems to increase stand-off range and to engage targets further inland [37]. The primary driver for the use of new ETPE-based propellants has been their excellent performance coupled with low flame temperatures [38]. Energetic solids (such as CL-20 and TEX) will provide additional burning rate and energy tailoring capability.

The flexibility of molding ETPE formulations into a wide range of geometries have been exploited in the new co-layered propellants in which, a relatively slower burning propellant is used in the outer core and a faster burning composition is used in the inner core. The inner-core propellant begins to burn as the propellant moves down the bore, which helps to maintain a high level of pressure for a relatively longer duration [39] (Figure 4.17).

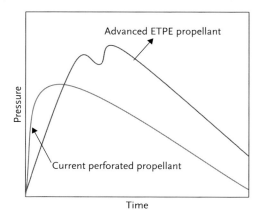

Figure 4.17 Schematic comparing the pressure–time traces for advanced ETPE propellant and perforated propellant.

Table 4.2 Comparison of predicted performance parameters and experimental sensitivity properties of ETPE propellants with NC-based baseline propellants [40].

Propellant	RP-36	RP-1315	TGD-043	TGD-044
Binder	NC	NC	BAMO-GAP	BAMO-AMMO
Density (g/cm^3)	1.5871	1.6290	1.5920	1.5901
Force constant (J/g)	926	999	1177	1175
Flame temperature (K)	2506	2888	2800	2800
Ballistic energy (J/g)	3502	4067	4259	4268
ABL impact (cm)	13	6.9	21	33
ABL friction (kN at 2.44m/s)	3.56	3.56	3.56	3.56
ESD (J)	>8	>8	>8	>8
SBAT (°C)	123.9	120.6	152.8	156.1

The calculated performance parameters of advanced ETPE layered propellants are compared with typical NC-based compositions in Table 4.2. Theoretically, the ETPE-based propellants provide higher impetus and ballistic energy at similar flame temperatures [40]. Furthermore, the new ETPE propellants are found to be relatively insensitive to initiation via friction, impact, and thermal and electrostatic stimuli. ETPE propellants are more thermally stable than the conventional double base formulations.

In the field evaluation trials, the ETPE propellants were observed to have a lower chamber pressure (202 MPa), which makes them attractive alternatives to the NC-based propellant that produces a chamber pressure of 365 MPa.

Luman et al. [41] used a "materials by design" approach to develop BAMO-AMMO ETPE based high performance layered propellants containing high energy

Table 4.3 Formulation and selected properties of ETPE layered propellant formulations [41].

Formulation	Constituents	Impetus (J/g)	Flame temperature (K)
A (high burn rate)	HNF/RDX/BAMO-AMMO/Al	1314	3432
B (low burn rate)	RDX/BAMO-AMMO/Al	1254	3224
75%A + 25%B (layered)	–	1299	3380

ingredients (HNF) and also nano-sized aluminum particles. The melt-casting possibility of ETPE formulations facilitates the development of layered propellant formulations with tailored burn rates, by combining propellants with faster and slower burn rates in adequate proportions (Table 4.3). The faster burn rate propellants were obtained by adding nanoscale aluminum to propellants with an oxidizer having a positive oxygen balance, such as HNF.

Thus ETPE-based formulations offer the exclusive advantage of flexibility to tailor the burn rates and consequently the performance of formulations by the melt-casting process, which the conventional polymer binder formulations do not provide. Advanced propellants with an average impetus of 1299 J/g and a flame temperature of 3380 K could be formulated, which are intended for fast-core layered propellant applications. The formulations have a burn-rate ratio of around three between the fast and slow burning layers [41].

4.6
Melt-Cast Explosives Based on ETPEs

Conventional melt-cast explosives such as Comp B or octol consist of TNT (trinitrotoluene) or a dispersion of high-energy oxidizers in TNT. These compositions are melted and cast into artillery shells, rockets, and bombs. However, these formulations exhibit poor mechanical properties and show undesirable defects, such as cracks and voids. Polymer binders have been introduced to the explosive formulations, which serve to improve the mechanical properties and also provide insensitive characteristics to the compositions.

While a number of PBX (polymer bonded explosive) systems processed by the cure-cast methods have been shown to provide improved sensitivity, the manufacture is typically more expensive than melt-cast options. Hence, weapon manufacturing facilities all over the world are geared towards melt-cast options, and research into new low sensitivity melt-cast options is continuing [42].

ETPE based PBX formulations, which could be processed by melt-casting techniques, are promising alternatives to the conventional cure-cast PBX. The main hindrance towards using thermoplastic elastomers for PBX formulations is that there are only a few thermoplastic elastomers that melt in the range of 80–100 °C, which is the desired temperature for melt-casting operations [43]. For example, in the case of GAP-based copolyurethane elastomers, the decomposition and melting occur at around 220 °C [43].

Table 4.4 Performance and shock sensitivity results of nitramine/ETPE-TNT melt-cast formulations [44].

Formulation	Density (kg/m³) (measured)	Detonation velocity (m/s) (measured)	P_{CJ} (GPa) (calculated)	Shock sensitivity (number of cards)
RDX/TNT/ETPE (75 : 17.5 : 7.5)	1.7	8107	27.9	203–204
HMX/TNT/ETPE (70 : 22.5 : 7.5)	1.76	8160	29.3	171–172
HMX/TNT/ETPE (69.5 : 20.5 : 10)	1.73	8064	28.1	167
RDX/TNT (60 : 40)	1.69	7885	26.3	216

In order to utilize these ETPEs in melt-cast systems, Ampleman et al. [43] at DREV Valcartier patented a process to manufacture insensitive melt-cast explosive compositions by dissolving an ETPE in melted TNT. The resulting ETPE/TNT solution is further mixed with other constituents of the explosive composition, such as nitramines [44], metal fuels, and plasticizers, which could be processed with the melt-cast facilities. In order to lower the viscosity of the mixes, the polymerization of ETPE is carried out *in situ* during the mixing and casting of the formulation.

Melt-cast formulations based on RDX and HMX in TNT/ETPE melt have exhibited better performance and lower sensitivity to external stimuli than Comp B (RDX in TNT melt) (Table 4.4).

The velocity of detonation of ETPE formulations showed an increase of up to 108% compared with Comp B and decreases with increasing percentage of ETPE in the formulation. The ETPE formulations with higher percentages of nitramines are less shock sensitive than Comp B itself. Table 4.4 exemplifies the effectiveness of ETPE in producing insensitive melt-cast formulations with higher performance.

4.7
ETPE Based Polymer Nanocomposites

Polymers reinforced by discrete constituents of the order of a few nanometers are referred to as polymer nanostructured materials or polymer nanocomposites (PNC). Uniform dispersion of nanoscale particles or nanoelements in the polymer matrix provides immense interfacial areas (of the order of 700 m²/cm³). Moreover, the distance between the nanoelements dispersed in the polymer is of the order of 10 nm. These two features differentiate PNCs from the conventional filled plastics. Other than providing mechanical reinforcements of the base polymer, the value of PNC technology comes from providing value-added properties that are not present in the base resin without sacrificing its inherent mechanical properties or processability [45].

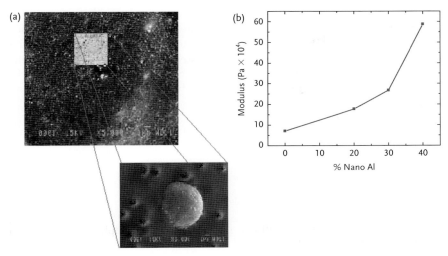

Figure 4.18 (a) Scanning electron micrograph of ETPE/Al composite with 40% micrometric aluminum content. (b) Effect of addition of nanometric Al on the storage modulus of ETPE/Al nanocomposite. Reproduced with permission of Wiley from [47].

Layered silicates are the most commonly used inorganic nanoelements in PNC research. Nanoscale dispersion of the layered silicates in resins provides significant improvements in glassy and rubbery moduli. Additionally, it offers supplementary benefits of enhanced thermal stability, reduced thermal expansivity, flammability, matrix swellability, and gas permeability, which could be utilized for the applicability of PNCs for defence applications [46].

For improving the mechanical properties of energetic polymers such as ETPE, nanoscale aluminum is a better choice as a reinforcement because it is a common ingredient in the energetic formulations as a fuel. Diaz et al. [47] explored the application of nanometric aluminum as a reinforcement for ETPE based on GAP copolyurethane.

An SEM micrograph of the cross-section of the nanocomposite shows uniform dispersion of the particles and good interfacial adhesion between the particle and the polymer matrix, producing an enhancement in the storage modulus (Figure 4.18).

Interestingly, the T_g of ETPE ($-40\,°C$) does not vary with the addition of aluminum, which means there is no molecular level interaction between the reinforcement and the polymer matrix. The densities of the nanocomposite were found to increase with increasing aluminum content, indicating that the energetic performance parameters could be enhanced by the inclusion of aluminum nanoparticles. Denser materials provide higher detonation velocities and pressures. The impact sensitivity of the nanocomposites decreases with increasing amount of aluminum.

There is future scope for developing ETPE/Al composites as reactive nanomaterials if there is a sufficient amount of oxidizing agent to release the volumetric heat of oxidation from the aluminum metal. The oxidizer could be any transition metal oxide such as iron oxide or molybdenum oxide, so that the Al/metal oxide system can be regarded as a nanothermite. Advancements in macromolecular chemistry have provided synthetic routes to assemble the metal around the metal oxide to maximize the rate of energy release.

References

1. Holden, G. (2000) *Understanding Thermoplastic Elastomers*, Hanser, Munich.
2. Spontak, R.J. and Patel, N.P. (2000) Thermoplastic elastomers: fundamentals and applications. *Curr. Opin. Colloid Interface Sci.*, **5**, 334–341.
3. Hamley, I.W. (1998) *The Physics of Block Copolymers*, Oxford University Press, Oxford.
4. Johnson, N.C., Gill, R.C., Leahy, J.F., Gotzmer, C., and Fillman, H.T. (1990) Melt cast thermoplastic elastomeric plastic bonded explosive. US Patent 4,978,482.
5. (a) Manser, G.E. and Ross, D.L. (1982) Energetic thermoplastic elastomers. Report of SRI International Prepared for Office of Naval Research, report No: AD A122909, Office of Naval Research; (b) Ampleman, G., Brochu, S., and Desjardins, M. (2001) Synthesis of energetic polyester homopolymers and ETPEs formed therefrom. Defence Research Establishment Valcartier Technical Report, report No: 2001–175, Defence Reserch Establishmnet Valcartier.
6. Beaupréet, F., Ampleman, G., Nicole, C., and Melancon, J.G. (2003) Insensitive propellant formulations containing energetic thermoplastic elastomers. US Patent 6,508,894.
7. Wardle, R.B., Edwards, W.W., and Hinshaw, J.C. (1996) Methods of producing thermoplastic elastomers having alternate crystalline structure such as polyoxetane ABA or star block copolymers by a block linking process. US Patent 5,516,854.
8. Sanderson, A.J., Edwards, W., Cannizzo, L.F., and Wardle, R.B. (2006) Synthesis of energetic thermoplastic elastomers containing both polyoxirane and polyoxetane blocks. US Patent US 7,101,955 B1.
9. Xu, B., Lin, Y.G., and Chien, J.C.W. (1992) Energetic ABA and (AB)$_n$ thermoplastic elastomers. *J. Appl. Polym. Sci.*, **46**, 1603–1611.
10. Sanderson, A.J. and Edwards, W. (2004) Synthesis of energetic thermoplastic elastomers containing urethane linkages. US Patent 6,815,522.
11. Wardle, R.B., Cannizzo, L.F., Hamilton, R.S., and Hinshaw, J.C. (1992) Energetic oxetane thermoplastic elastomer binders. Final Report to Office of Naval Research from Thiokol Corporation Under Contract Number N00014-90-C-0264, Thiokol Corporation.
12. Kimura, E., Oyumi, Y., Kawasaki, H., Maeda, Y., and Anan, T. (1994) Characterization of BAMO/NIMMO copolymers. *Propellants Explos. Pyrotech.*, **19**, 270–275.
13. Pisharath, S. and Ang, H.G. (2007) Synthesis and thermal decomposition of GAP-Poly(BAMO) copolymer. *Polym. Degrad. Stab.*, **92**, 1365–1377.
14. Kawamoto, A.M., Barbieri, U., Polacco, G., Krause, H., Sabio Holanda, J.A., Keiser, M., and Keicher, T. (2008) Synthesis and characterization of random copolymers of GAP and Poly (BAMO). *Propellants Explos. Pyrotech.*, **33** (5), 365–372.
15. Talukder, M.A.H. and Lindsay, G.A. (1990) Synthesis and preliminary analysis of block copolymers of BAMO and NIMMO. *J. Polym. Sci. A: Polym. Chem.*, **28**, 2393–2401.
16. Liu, Y.-L., Hsiue, G.-H., and Chiu, Y.-S. (1995) Studies on the polymerization mechanism of NIMMO and AMMO and the synthesis of their respective

copolymers with THF. *J. Polym. Sci. A: Polym. Chem.*, **33**, 1607–1613.
17 Al-Kaabi, K. and Van Reenen, A. (2007) Synthesis of energetic thermoplastic elastomers using controlled radical polymerization. Proceedings of Insensitive Munitions & Energetic Materials Technology Symposium, Miami, USA.
18 Piraino, S., Kaste, P., Snyder, J. Newberry, J., and Pesce-Rodriguez, R. (2004) Chemical and structural characterization of energetic thermoplastic elastomers: BAMO/AMMO copolymers. Proceedings of 35th International Annual Conference of ICT, Karlsruhe, Germany.
19 Kimura, E. and Oyumi, Y. (1995) Thermal decomposition of BAMO copolymers. *Propellants Explos. Pyrotech.*, **20**, 322–326.
20 Liu, Y.L., Hsiue, G.H., and Chiu, Y.S. (1995) Thermal characteristics of energetic polymers based on tetrahydrofuran and oxetane derivatives. *J. Appl. Polym. Sci.*, **58**, 579–586.
21 Chang, T.C., Wu, K.H., Chen, H.B., Ho, S.Y., and Chu, Y.S. (1996) Thermal degradation of aged polytetrahydrofuran and its copolymers with 3-azidomethyl-3′-methyloxetane and 3-nitratomethyl-3′-methyloxetane by thermogravimetry. *J. Polym. Sci. A Polym. Chem.*, **34**, 3337–3343.
22 Pisharath, S. and Ang, H.G. (2007) Synthesis and thermal degradation of GAP-Poly(BAMO) copolymer. *Polym. Degrad. Stab.*, **92**, 1365–1377.
23 Vyazovkin, S. and Dollimore, D. (1996) Linear and nonlinear procedures in isoconversional computations of activation energy of nonisothermal reactions in solids. *J. Chem. Inf. Comput. Sci.*, **36**, 42–45.
24 Kimura, E. and Oyumi, Y. (1995) Thermal decomposition of BAMO copolymers. *Propellants Explos. Pyrotech.*, **20**, 322–326.
25 Shlensky, O.F., Vaynsteyn, E.F., and Matyukhin, A.A. (1988) Dynamic thermal decomposition of linear polymers and its study by thermo analytical methods. *J. Therm. Anal.*, **34**, 645–655.
26 Vyazovkin, S. and Wight, C.A. (1997) Kinetics in solids. *Annu. Rev. Phys. Chem.*, **48**, 125–149.
27 Ling, P. and Wight, C.A. (1997) Laser photodissociation and thermal pyrolysis of energetic polymers. *J. Phys. Chem. B.*, **101**, 2126–2131.
28 Eroglu, M.S. and Guven, O. (1996) Thermal decomposition of poly (glycidyl azide) as studied by high temperature FTIR and thermogravimetry. *J. Appl. Polym. Sci.*, **61**, 201–206.
29 Sell, T., Vyazovkin, S., and Wight, C.A. (1999) Thermal decomposition kinetics of PBAN binder and composite solid propellants. *Combust. Flame*, **119**, 174–181.
30 Vyazovkin, S. and Wight, C.A. (1999) Kinetics of thermal decomposition of cubic ammonium perchlorate. *Chem. Mater.*, **11**, 3386–3393.
31 Lee, Y.J. and Litzinger, T.A. (2002) Thermal decomposition of BAMO/AMMO with and without TiO_2. *Thermochim. Acta*, **384**, 121–135.
32 Lee, Y.J., Kudva, G., and Litzinger, T.A. (1999) Thermal decomposition of BAMO/AMMO copolymer. Proceedings of 35th AIAA Joint Propulsion Conference and Exhibit, Los Angeles, USA.
33 Braithwhite, P., Edwards, W., Sanderson, A.J., and Wardle, R.B. (2001) The synthesis and combustion of high energy thermoplastic elastomer binders. Proceedings of the 32nd International Annual Conference of ICT, Karlsruhe, Germany, 9.1–9.7.
34 Kimura, E., and Oyumi, Y. (1996) Insensitive munitions and combustion characteristics of BAMO/NIMMO propellants. *J. Energ. Mater.*, **14**, 201–215.
35 (a) Zenin, A. and Finjakov, S. (2004) Physics of combustion of solid mixtures of active binders with new oxidizers. Proceedings of the 35th International Annual Conference of ICT, Karlsruhe, Germany, 144.1–144.16; (b) Zenin, A. and Finjakov, S. (2002) Physics of combustion of energetic binder–nitramine mixtures. Proceedings of the 33rd International Annual Conference

of ICT, Karlsruhe, Germany, 6.1–6.4; (c) Zenin, A. and Finjakov, S.V. (2001) Physics of combustion of HTPB/ nitramine compositions. Proceedings of the 32nd International Annual Conference of ICT, Karlsruhe, Germany, 8.1–8.2.

36 Paletsky, A.A., Volkov, E.N., Korobeinichev, O.P., and Tereshchenko, A.G. (2007) Flame structure of pseudo-propellants based on nitramines and azide polymers at high pressure. *Proc. Combust. Inst.*, **31**, 2079–2087.

37 Navy Mantech Manufacturing Technology Program on Continuous Extrusion Safety and Efficiency of Propellant Manufacturing. S0984-Flexible Manufacturing of Nitrogen Based Gun Propellants, Naval Surface Warfare Centre, Indian Head Division, March 2010.

38 Braithwaite, P., Dixon, G., Rose, M., and Wardle, R. (2002) The promise of energetic TPE gun propellants – from notebook to full scale verification. Proceedings of NDIA Annual Gun and Ammunition Symposium. Panama City, Florida, USA.

39 Committee on Advanced Energetic Materials and Manufacturing Technologies (2004) Advanced gun propellants, in *Advanced Energetic Materials*, The National Academies Press, Washington, DC.

40 Manning, T., Cramer, M., Ray, M., and Braithwaite, P. (2006) Performance of co-layered ETPE propellant in medium caliber ammunition. Proceedings of Insensitive Munition and Energetic Material Symposium. Bristol, UK.

41 Luman, J.R., Wehrman, B., Kuo, K.K., Yetter, R.A., Masoud, N.M., Manning, T.G., Harris, L.E., and Bruck, H.A. (2007) Development and characterization of high performance solid propellants containing nano-sized ingredients. *Proc. Combust. Inst.*, **31**, 2089–96.

42 Provatas, A. and Davies, P.J. (2006) Australian melt cast explosives R&D-DNAN a replacement for TNT in melt cast formulations. Proceedings of Insensitive Muntions and Energetic Materials Symposium, Bristol, UK.

43 Ampleman, G., Brousseau, P., Thiboutot, S., Dubois, C., and Diaz, E. (2003) Insensitive melt cast compositions containing energetic thermoplastic elastomers. US Patent 6,562,159.

44 (a) Brousseau, P., Ampleman, G., Thiboutot, S., Diaz, E., Trudel, S., Beland, P., Duval, D., Blanchet, J.F., Gosselin, P., and Judge, M.D. (2006) High performance melt cast plastic bonded explosives. Proceedings of the NDIA Insensitive Munitions and Energetic Material Symposium, Bristol, UK; (b) Brousseau, P., Ampleman, G., Thiboutot, S., and Diaz, E. (2001) New melt cast explosives based on energetic thermoplastic elastomers. Proceedings of 32nd International Annual Conference of ICT on Energetic Materials, Karlsruhe, 89/1–14; (c) Thiboutot, S., Brousseau, P., Ampleman, G., and Pantea, D. (2008) Potential use of CL-20 in TNT/ETPE based melt cast formulations. *Propellants Explos. Pyrotech.*, **33** (2), 103–108.

45 Paul, D.R. and Robeson, L.M. (2008) Polymer nanotechnology: Nanocomposites. *Polymer*, **49**, 3187–3204.

46 Vaia, R. (2002) Polymer nanocomposites open a new dimension for plastics and composites. *AMPTIAC Quart. Newslett.*

47 Diaz, E., Brousseau, P., Ampleman, G., and Prud'homme, R.E. (2003) Polymer nanocomposites from energetic thermoplastic elastomers and Alex®. *Propellants Explos. Pyrotech.*, **28** (4), 210.

5
Fluoropolymers as Binders

5.1
Introduction

Fluoropolymers contain highly electronegative fluorine atoms in the polymer chain. The electronegative nature of the fluorine atom makes it a strong oxidizer, which releases a significant amount of heat energy especially in the presence of metals. For example, two-component explosives containing metal particle fuels and a fluoropolymer oxidizer are able to produce more than twice as much as high performance molecular explosives, such as Her Majesty's Explosive (HMX) [1]. The inclusion of the fluoropolymers aids in enhancing the combustion efficiency of metallic particles due to the presence of Fluorine [2]. In addition to their oxidizing nature, they are highly dense (a close match with the density of high-explosive crystals), chemically inert, and exhibit robust thermal stability over large temperature ranges.

Among the fluoropolymers, poly(tetrafluoroethylene) (PTFE) is the most common, which is well known under the Dupont brand name of Teflon. It has been used as a component of reactive nanomaterials and as a binder in simple energetic material formulations. However, PTFE has several disadvantages from the perspective of formulations, which include high melt viscosity and melt temperature, and lack of solubility in common organic solvents [3]. Copolymerization of tetrafluoroethylene dramatically improves the physical properties of the polymer while maintaining high density and chemical inertness [3]. Examples include Viton and Kel-F800, which are widely used as binders for insensitive high explosives.

Polymers containing difluoroamino groups ($-NF_2$) have been extensively studied as ingredients for propellants and explosives due to their higher energy and positive heat of formation compared with other frequently used pendant energetic groups [4].

This chapter provides an overview and state-of-the art of the different fluoropolymers that are used as binders for propellant and explosive applications.

Energetic Polymers: Binders and Plasticizers for Enhancing Performance, First Edition.
How Ghee Ang and Sreekumar Pisharath.
© 2012 WILEY-VCH Verlag GmbH & Co. KGaA, Weinheim. Published 2012 by WILEY-VCH Verlag GmbH & Co. KGaA

5.2
Poly(tetrafluoroethylene) (PTFE)

5.2.1
Phase Transitions of PTFE

PTFE is the most recognized polymer in the fluoropolymer family. The original PTFE resin was accidentally discovered by Roy J. Plunkett in 1938, when he was trying to make a CFC (chlorofluorocarbon) refrigerant. The per(fluoroethylene) polymerized to form PTFE in a pressurized storage container, with the iron inside the container acting as a catalyst. The powdered PTFE as produced by the polymerization is a chain polymer, with a molecular weight of the order of half a million. PTFE is highly crystalline with the first-order transition from the crystalline to the amorphous state taking place at 330 °C [5]. The properties of PTFE are strongly dependent on the degree of crystallinity, which can vary over a wider range.

PTFE exists in four phases depending upon pressure and temperature [6(a)] (Figure 5.1). It exhibits two crystalline phase transitions at 19 and 30 °C at atmospheric pressure, which are unique among polymers. Below 19 °C, PTFE exists as phase II consisting of linear PTFE chains in a helical conformation containing 13 CF_2 groups per 180° turn [5, 6].

The transition at 19 °C involves a 1% change in density ascribed to the partial disordering of the crystalline regions. Between 19 °C $< T <$ 30 °C, PTFE exists as a more ordered hexagonal phase IV containing 15 atoms/180 °C with an associated increase in hexagonal lattice spacing [5, 6].

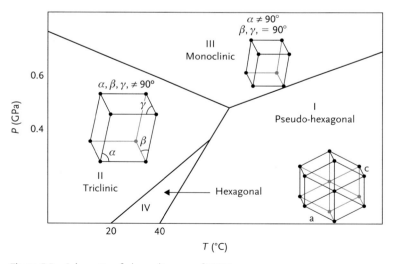

Figure 5.1 Schematic of phase diagram of PTFE.

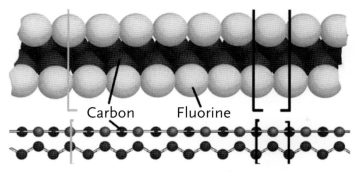

Figure 5.2 Crystalline structure of PTFE in phase III. Reprinted with permission from American Institute of Physics Copyright (2007).

At $T > 30\,°C$, further rotational disordering and untwisting of helices occurs, giving way to phase I of PTFE. Phase I exists as a pseudo-hexagonal structure, in which the individual polymer chains lose their well defined helical repeating unit [5, 6].

At elevated pressures (above 0.5 GPa) PTFE undergoes a phase change to planar zig-zag form in phase III [7–9] (Figure 5.2).

The phase III structure is reported to have a monoclinic structure with two molecules per unit cell [7]. This transition has been experimentally observed under shock-wave loading with traditional plate impact methods [9] and also by laser shock compression studies [10]. The phase results in a 13% local volume change within the crystalline domains and a considerable reduction in compressibility [7]. Shock prestraining from phase II to phase III results in increased crystallinity (which is independent of pressure) and enhancement in yield stress and Young's modulus, both parameters increasing with pressure.

5.2.2
Energetic Composites of PTFE

PTFE composites with Al have been classified as reactive materials, which are composites of inert solid materials when subjected to violent mechanical stimulus such as an impact, and react exothermally with rapid release of energy [11]. The reaction of Al with PTFE generates 21 GJ/m^3 compared with the best molecular explosive that generates less than 12 GJ/m^3 [12] Eq. (5.1).

$$4Al + 3(C_2F_4) \rightarrow 4AlF_3 + 6C \quad Q = 8670\,kJ/kg \tag{5.1}$$

PTFE must be first degraded for the fluorine to be available for reaction with Al fuel. The decomposition of PTFE starts at 803 K and completes around 900 K. During this decomposition an exothermic gasification reaction occurs and fluorine is produced in abundance [13]. Osborne and Pantoya [14] demonstrated that the reaction between Al and fluorine depends on the particle size of Al. With the micrometer size Al particles, when subjected to slow heating conditions, much of the fluorine produced during thermal decomposition of PTFE escaped the reaction zone before reacting with the micrometer size Al particles. On the other hand, when the same experiments were conducted using nanoscale Al, almost 75% of the PTFE reacted with Al, resulting in lower ignition temperatures and larger exothermic activity for the nanometer Al/Teflon mixture. This is caused by the pre-ignition reaction that is unique to the nano-Al/PTFE mixture controlled by the fluorination of the Al particle passivation shell [14].

Dolgoborodov and coworkers used the concept of mechanochemical activation to prepare mechanically activated energetic composites of Al and PTFE and studied its detonation and combustion behavior [15]. The mean particle size of Al particles used in the experiment was 3.6 µm. The mechanochemical activation was carried out in a vibratory mill and conditions were chosen in such a way that maximum homogenization of the reactant mixture was achieved in the absence of any chemical reaction. The unique feature of the detonation of the Al/PTFE composite is that, unlike the case with homogeneous explosives, the initial and final products of detonation are in a solid state. It is presumed that in such systems the reactions occur in the centers of the contact surface of the reagent particles, rather than in the entire volume of the material [15(a)]. Other than the chemical characteristics of the reagents, the reaction rate is also determined by the time of component mixing at the shock wave front [15(a)]. The detonation experiments were carried out in thick-wall composite tubes containing the Al/PTFE composite fitted with electric contact gauges and optical fibers for velocity measurement [15(a)].

Interestingly, the detonation velocity of Al/PTFE composite depends on the concentration of aluminum (Figure 5.3). The composition containing a 25% Al mixture at 0.54 g/cm^3 recorded a detonation velocity of 1280 m/s. The ability of the Al/Teflon mixture to detonate is important for the explosive applications.

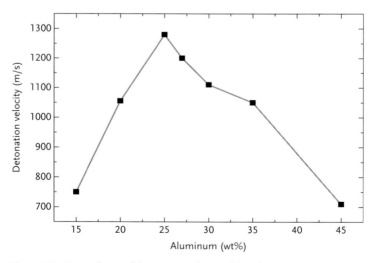

Figure 5.3 Dependence of detonation velocity of the Al/PTFE composites on the Al content. Values for the plot are taken with permission of Springer from Ref. [15(b)].

It is concluded that the detonation of Al/PTFE composite is controlled by a relay mechanism in which jets of the reaction products originating from one hot spot initiates the reaction in other hot spots [15(b)]. The results demonstrated the application of mechanochemical activation on enhancing the reactivity of Al/PTFE composites and, under shock initiation, the possibility of reaching the steady state detonation regime with a detonation velocity of 1300 m/s.

Burning behavior of Al/PTFE composites were studied through igniting the mixture by an electrically heated Nichrome wire and measuring the burning progress with the help of an optical fiber. The burning velocity ranges from 160 to 210 m/s for mechanochemical activated Al/PTFE composites [15(c)]. This burn rate was found to be lower than that recorded for Al/MoO$_3$ thermite mixtures, where a self-accelerating chemical reaction (average burn velocity of 400 m/s) with a transition from combustion to explosion has been observed.

Watson *et al.* [16] compared the combustion behaviors of Al/PTFE and Al/MoO$_3$ composites in confined and unconfined environments. The behaviors were interpreted in terms of flame propagation speeds measured by high-speed imaging.

In confined environments, the gases generated as a result of the combustion reaction cannot escape to any significant extent and enhance the convective mode of energy propagation. Hence more fluorine gas is available for reaction with Al because the gas is unable to escape from the reaction system [16]. On the other hand, for unconfined burning environments, the fluorine gas produced escapes to the surroundings and does not contribute to the convective mode of energy propagation [16].

As illustrated in Figure 5.4, the Al/PTFE composite containing 40% of nano-Al (50 nm) provides a flame propagation speed of 840 m/s. However, the composite

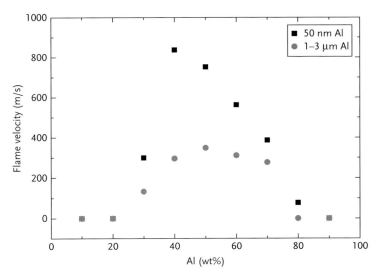

Figure 5.4 Dependence of confined flame propagation velocity of Al/PTFE composites on Al content with different particle sizes. Values for the plot are taken with permission of Elsevier from Ref. [16].

containing the same amount of micron-sized Al (1–3 μm) provides a flame propagation speed of 295 m/s only [16], demonstrating the difference in reactivity between the nano- and micron-sized Al towards PTFE. The flame propagation speed of nano-Al/PTFE composites is observed to be lower than that for nano-Al/MoO_3 composites, even though the former provides a higher peak pressure due to fluorine gas generation [16]. Thus, no direct correlation exists between the generated gas pressures and flame speeds in nano-Al/oxidizer thermite systems. It is concluded that the flame propagation speed is limited by the efficiency of the collisions between Al bare clusters dispersed by the melt-dispersion mechanism [17] and solid or gaseous oxidizer or gaseous fluorine.

Hunt et al. investigated [18] the impact ignition of nano-Al/PTFE composites using a modified type-12 impact tester. The ignition energies were found to be of the order of 4 J for Al/PTFE composites, with the micron-scale Al particles requiring higher energy than the nanoscale particles. Impact tests on Al mixed with Ni and MoO_3 resulted in lower ignition energies ranging between 0.2 and 1.7 J and generally increased with increasing particle size. The higher impact ignition energies for Al/PTFE composites have been attributed to the cushioning effect of PTFE on the Al particles due to its high hardness and low ductility [18]. It is also noted that the ignition mechanism of impact-induced initiation is diffusion-controlled due to the rupture of the shell structure [19], in contrast to the melt-dispersion mechanism operating in laser-induced initiation.

A prospective application of Al/PTFE reactive materials are as fragmentation components into munitions. Reactive materials provide not only destruction similar to that achieved with inert fragments but also energy release after penetration [20]. Experimental firings of reactive material based penetrators against soft targets have proved the efficiency of reactive materials in causing extensive external damage as compared with conventional fragment performance [21, 22].

5.3
Copolymers of Tetrafluoroethylene

The main drawbacks in the application of PTFE as an explosive binder are its high melting-point (340 °C) and viscosity (10^9–10^{12} Pa s) and also the insolubility in common organic solvents. In order to overcome these shortcomings, TFE is often copolymerized with other fluorinated monomers. Copolymerization influences the crystallinity and phase behavior of TFE. When copolymerized with hexafluoropropylene (HFP), the crystallization tendency of TFE is reduced by introducing sterically large $-CF_3$ groups [23]. The crystallinity of the copolymer may vary between 70% in virgin polymer and 30–50% for the molded parts, depending on the processing conditions. The melting-point is reduced from the original of 325 °C to about 260 °C. The side groups also inhibit the chain slippage thus reducing the creep [23]. In contrast to the case with PTFE, the copolymer exhibits only a single melting-point [24].

Two important copolymers of TFE that are used in explosive binder applications are Kel-F800 and Viton B.

5.3.1
Kel-F800

Kel-F800, developed by 3M, is a thermoplastic fluorocarbon polymer widely used as a binder in PBX formulations [25].

It is a random copolymer of chlorotrifluoroethylene (CTFE) and vinylidene fluoride (VDF) monomers in a 3 : 1 ratio (Figure 5.5). The presence of vinylidene fluoride disrupts the crystallinity of chlorotrifluoroethylene to form an amorphous polymer with a softening temperature of 105 °C [26]. The polymer is very dense (density = 2.120 g/cm^3) due to the presence of the chlorine and fluorine atoms and T_g is observed between 28 and 38 °C [26(a)]. The linear copolymer is soluble in several common organic solvents, such as acetone, methyl methyl ketone, and THF [26]. 3M phased out the application of perfluorooctanoyl derivatives used as an emulsifier for the manufacture of Kel-F800 due to bioaccumulation issues. This led to the discontinuation of the manufacture of the polymer. Instead a new polymer, FK-800, which is indistinguishable from Kel-F800, has been introduced. It has been experimentally proved as a feasible candidate for replacement of Kel-F800 in current insensitive high explosive compositions such as LX-17 and PBX 9502.

5.3.1.1 Dynamic Behavior of Kel-F800 under Shock
The role of polymer binders in explosive formulations is to improve the mechanical properties and desensitize the explosive from the influence of external shocks. During the service period of the PBX, the composites are stored for longer periods under conditions of cyclically varying temperatures and also subjected to unplanned mechanical events. Hence, understanding and modeling the stress–strain response of explosive polymer binders such as Kel-F800 as a function of strain rates and temperature has gained research interest.

Blumenthal et al. [27] investigated the influence of strain rate and temperature on the mechanical behavior under compression of Kel-F800. It was observed that Kel-F800 is much less strain-rate and temperature dependent than the polyurethane

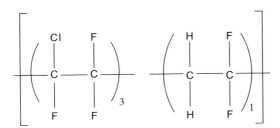

Figure 5.5 Chemical structure of Kel-F800.

based explosive binder Estane [27]. High strain rate loading of Kel-F800 increases the temperature dependence of the flow stress. It also shifts the flow stress versus temperature profile to higher stresses. The high strain rate compressive stress–strain profile of Kel-F800 is compared with that of PTFE and Estane in Figure 5.6 [28].

Kel-F800 is observed to be five times stronger in compression than PTFE and ten times stronger than pure Estane. The Kel-F800 also exhibits a peak in the flow stress at 10% strain in contrast to PTFE or Estane, showing essentially linear stress–strain responses, under both quasistatic and dynamic conditions [27, 28]. The decrease in flow stress for Kel-F800 after the peak demonstrates the onset of crack damage in the samples under the influence of uniaxial compressive stress [28].

Bourne and Gray [28] investigated the shock induced damage of Teflon and Kel-F800 in terms of the spall strengths and lateral stresses as a function of the impact stress in the range of (0.6–3.5 GPa). Spall or dynamic fracture of the material occurs when the tensile stresses from the colliding rarefaction waves (caused by the reduction in density following a shock wave) exceed the strength of the material. The tensile spall strength was measured by the inelastic impact of a flyer plate accelerated from a gun to speeds of up to about 800 m/s onto a polymer target. The free surface of the target is set to motion by a shock roughly a microsecond after the impact. The shock waves generated by impact reflect at the outer surfaces of the flyer and target, generating pressure reducing waves known as rarefactions. Rarefactons collide near the middle of the target, and the generated tension is sufficient to cause the material to spall [29]. The velocity of the free

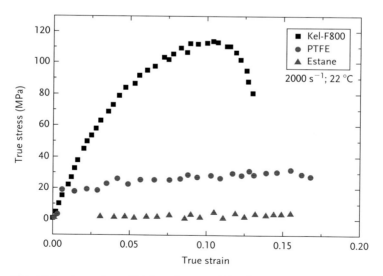

Figure 5.6 Comparison of high strain rate (2000 s^{-1}) compressive behavior of pure Kel-F800 with Estane at 22 °C [28]. Reproduced with permission from American Institute of Physics, Copyright (2005).

surface for spall was measured by VISAR (velocity interferometer system for any reflector) [30]. It was observed that Kel-F800 shows decreasing spall strength with increasing impact stress, while PTFE displayed no measurable spall strength even at the lowest velocity tested [28].

In the second set of experiments to investigate the shock induced response, lateral stresses were measured by placing manganin gauges at a distance of 4 mm from the impact face [28]. Measurement of lateral stress facilitates the assessment of shear strength of the polymer, when combined with the longitudinal data deduced from known impact conditions through impedance matching techniques. Lateral stresses for Kel-F800 and PTFE has been observed to be constant behind the shock front. Kel-F800 shows constant shear strength with the shock amplitude, while for PTFE the shear strength increases with the magnitude of the impact stress [28]. This feature indicates that Kel-F800 is more prone to shock induced damage at higher stresses as compared with PTFE [28].

Shock velocity measurements are also used to generate the Hugoniot loci and equation of state (EOS) for several polymers including explosive binders [3]. The measurement data are often presented in the shock velocity (U_s)–particle velocity (U_p) frame [3]. Barring phase transitions and other dynamic processes, the relationship between U_s and U_p is expected to be linear; $U_s = C_0 + S U_p$. The Y-intercept will provide the bulk sound velocity (C), and the slope of the data (S) will provide the pressure derivative of isentropic bulk modulus [3]. All the important thermodynamic parameters related with the EOS of polymers could be derived from the knowledge of C_0 and S. In general, Carter and Marsh have observed that polymers universally undergo a high-pressure transition marked by a change in the U_s and U_p plane for $U_p \sim 2$–3 km/s [31].

Hugoniots have been generated for PTFE and Kel-F800 using shock velocity measurements for a range of pressures [25, 32–34].

As illustrated in Figure 5.7, the shock Hugoniot of PTFE exhibits a subtle cusp in the profile, at low pressures, due to the II–III phase transition associated with a volume change of 2.2%.

The linear fits to the data above and below the cusp are $U_s = 1.33 + 2.67 U_p$ at pressures <7 kbar and $U_s = 1.22 + 2.46 U_p$ at pressures >7 kbar [33].

For Kel-F800, the linear fit to the shock Hugoniot is represented by $U_s = 1.824 + 1.84 U_p$ [25]. All the Hugoniot data are represented in the linear fit for pressure ≥ 30 kbar, without any evidence of phase transitions. It is interesting to note that the shock Hugoniot of Kel-F800 is independent of crystallinity of the samples as well as on the nature of heat treatments the samples are subjected to. The compression-molded and melt-processed Kel-F800 behaves similarly (obeying the same Hugoniot) under the influence of shock despite a difference in crystallinity of 10–15% and a difference in density of 2% [3, 25].

5.3.1.2 Thermal Decomposition

PTFE when degraded in an inert atmosphere gives the tetrafluoroethylene (TFE) monomer in 100% yield. When burned in air, the significant products are TFE and COF_2 [35]. The thermal decomposition kinetics of Kel-F800 have been studied in

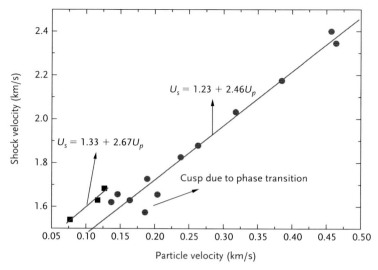

Figure 5.7 Shock Hugoniot of PTFE showing different linear fits for points above and below the cusp [33]. Reprinted with permission of American Institute of Physics, Copyright (2004).

the literature [36]. The decomposition of Kel-F800 was modeled by two parallel reactions: an early decomposition process accounting for 1% of the initial weight loss of the polymer followed by an autocatalytic reaction [36(d)]. No residue was obtained after the decomposition process. The activation energy of the thermal decomposition is in the range of 220–271 kJ/mol. It has been observed that high molecular weight products with a mean molecular weight of 490 Da are formed in significant amounts [36(a)]. At higher temperatures (>300 °C) hydrogen halides and fluorine are evolved [36(a)].

A free-radical mechanism has been proposed for the thermal decomposition of Kel-F800 [36(c)], which has been confirmed by the presence of free radicals in the electron spin resonance spectra. The polymer radicals formed during heating function as initiators and promote the scission of hydrogen halides from the fluorine- and chlorine-containing blocks and then from the polymer chain itself [36(d)]. This process occurs with activation energy of ∼125.6 kJ/mol. The scission of hydrogen halides is accompanied by the formation of conjugated double bonds along the polymer chain, which changes the color of the polymer chain. At higher temperatures (>300 °C) profound degradation of the polymer chain occurs releasing plenty of high molecular weight reaction products.

5.3.2
Viton A

Viton A is a random copolymer of vinylidene fluoride and hexafluoropropylene (Figure 5.8). The ratio of vinylidene fluoride (VDF) to hexafluoropropylene (HFP) repeat units is approximately 7 : 2. Viton A is an elastomer with a fluorine content

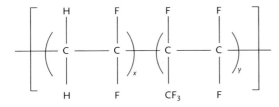

Figure 5.8 Chemical structure of Viton.

of 66% and a density of 1.78–1.82 g/cm³, ideal for an explosive binder. This fluorine content implies a 60 : 40 weight ratio of VDF to HFP monomers or approximately 77.83 mol% VDF. The glass transition temperature of Viton A is in the range of −18 to −25 °C [37].

The VDF-based fluoropolymers are usually cross-linked by four main agents: diamines, bisphenols, organic peroxides/triallyl isocyanurate systems, and by radiation. Whatever, the cross-linking agent, the mechanism of cross-linking requires two steps: the press cure and post cure. The press cure activates the curing mechanism to provide sufficient cross-links in the sample and the post-cure step is required to attain the best vulcanizate properties [38].

As an example, the curing of Viton with a diamine curing involves three steps [38] (Scheme 5.1):

i. dehydrofluorination from the VDF segments to generate internal double bonds;
ii. diamine curing agents form cross-links by reacting at the unsaturated cure sites;
iii. elimination of HF from the cross-links during the post cure to form further double bonds.

An acid acceptor of metal oxide (magnesium oxide) type is a necessary ingredient for the Viton curing formulations [38]. MgO contributes to the elimination of HF from the polymer during the curing reaction.

$$MgO + 2HF \rightarrow MgF_2 + H_2O \qquad (5.2)$$

The water formed as a result of the scavenging reaction of HF acts to inhibit the full development of cure unless it is removed from the vicinity of the polymer by post-curing in an open system [38(b)].

Peroxide-cure reactions proceed by a free-radical mechanism, which requires a special cure-site monomer along with the copolymer composition [38]. The common peroxide curing reagents used are dibenzoyl peroxide, dicumyl peroxide in association with coagents such as triallyl isocyanurate, which will enhance the cross-linking efficiency of peroxide curing.

As shown in Table 5.1, peroxide and amine cured vulcanizates differ in their mechanical properties.

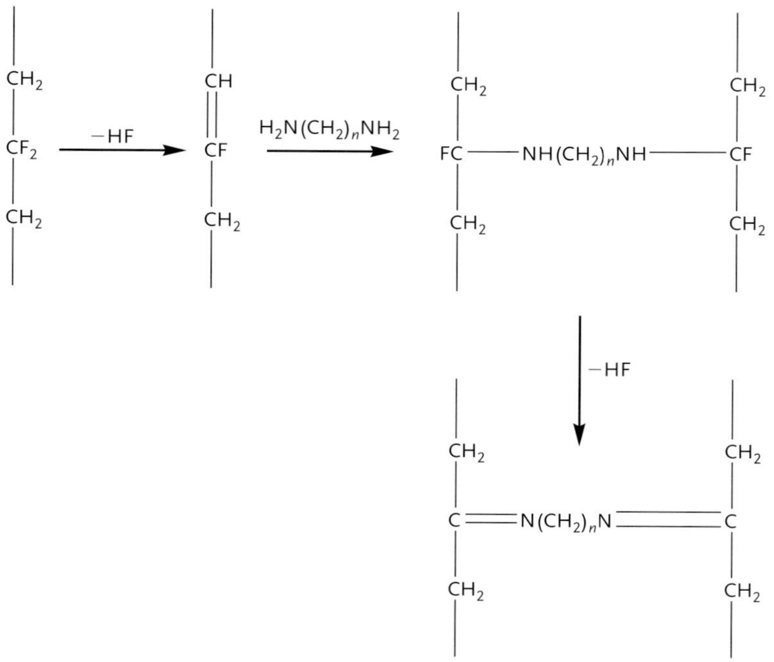

Scheme 5.1 Curing of Viton with diamine curing agent.

Table 5.1 Comparison of properties of Viton A vulcanizates cured by amine and peroxide [39].

Properties	Amine cure vulcanizate	Peroxide cure vulcanizate
Modulus at 100% (MPa)	8.14	2.48
Tensile strength (MPa)	18.27	15.51
Elongation at break (%)	180	720
Shore A hardness	75	75

The peroxide cure provides vulcanizates with better tensile strength and compression set resistance but lower modulus. On the other hand, the amine cured vulcanizates have higher modulus and lower elongation at break.

5.3.2.1 Thermal Decomposition of Viton

Thermal decomposition studied by TGA and DSC measurements indicates that Viton decomposes by a single-step process with a decomposition maximum at ∼450 °C. Knight et al. [40] studied thermal decomposition of cross-linked and uncross-linked Viton A in inert and oxidizing atmospheres by monitoring the evolution of breakdown products giving rise to F⁻ ions in aqueous solution by means of a fluoride ion-specific electrode. It was observed that the cross-linked sample of Viton A gave considerably more fluorine (29% average yield) in nitrogen than the uncross-linked analog, although there was little difference in air. This was assumed to be due to the formation of structures that favored dehydrofluorination in an inert atmosphere during the curing process. This result also hints at the lower thermal stability of the Viton gumstock brought about by the cross-linking process. No mechanistic interpretation of the decomposition reaction was provided.

Papazian [41] studied the degradation kinetics of Viton A at moderate temperatures using TGA measurements. The weight loss was described by a simple first-order equation and Arrhenius relationship plots were obtained from the TGA curves by calculating the effective rate constant at each conversion. Two activation energies were obtained from the break in the Arrhenius plot at 450 °C, at which point the activation energy changes from 106 to 356 kJ/mol.

Burnham and Weese [36(d)] applied the autocatalytic Prout-Tompkins extended model to the thermal decomposition of Viton by considering two parallel reactions. The calculated activation energies for the reactions are 167.2 and 216.8 kJ/mol, respectively. The thermal degradation of Viton produces ∼3% of residue, which is dependent on the heat rate.

5.4
Energetic Polymers Containing Fluorine

Compounds containing fluorine atoms bonded to nitrogen have been extensively studied as additives for propellant and explosive formulations. The attributes of the difluoroamino-groups are their capability to release larger, higher amounts of energy and positive heat of formation as compared with other frequently used energetic pendant groups such as the nitrato (–ONO_2) [4]. Furthermore, during the decomposition fluorine is liberated as the oxidizing agent, resulting in the improved metal combustion of solid propellants containing aluminum or boron metal fuels [42, 43]. Calculations have shown that –NF_2 compounds provide higher impulse values than their NO_2 counterparts [44].

Albeit the high predicted performance, practical application of the difluoroamino-compounds is limited by their unacceptably high impact sensitivity and instability. On the other hand, when the difluoroamino-group is placed at a neopentyl carbon, the compounds have remarkable thermal stability and low impact sensitivity [4]. The stability is provided by the steric hindrance.

Energetic mono- and bis-difluoroamino-monomers are prepared by the fluorination in nitrogen of the ethyl carbamate derivatives of the corresponding diamino- or amino-methyl oxetanes. The monomers are polymerized to form polyethers by ring opening polymerization with the help of Lewis acid catalysts (Scheme 5.2) in bulk at high temperatures. The hydroxyl terminated polyethers are subsequently cured to form energetic binder materials. Difluoroamine based

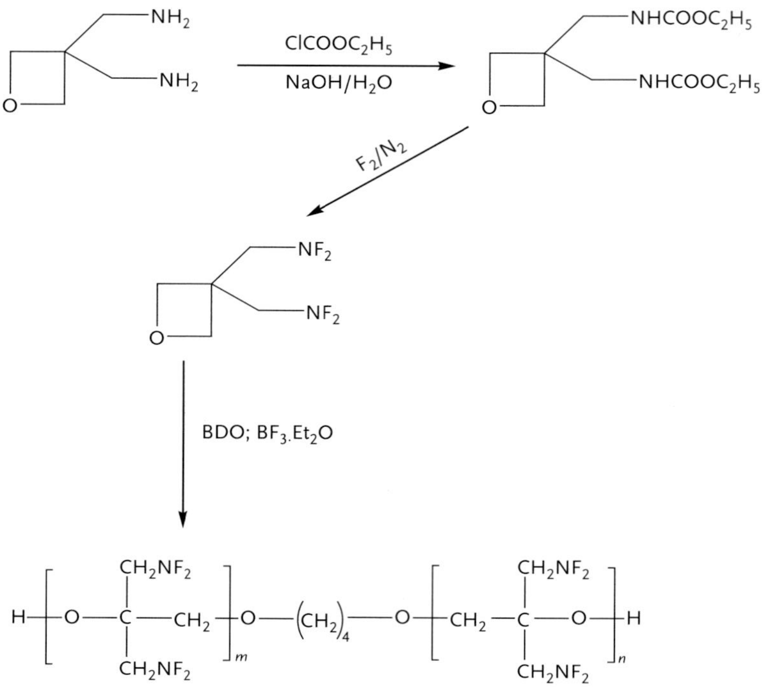

Scheme 5.2 Preparation of energetic bisdifluoroamine polymer [4].

energetic binders could also be prepared by the curing of 50:50 copolymer of mono- and bis-difluoroamine substituted oxetanes. Typical properties of the difluoramine polymer are presented in Table 5.2. It is a solid with melting-point of 158 °C at room temperature, which is not suitable for binder applications. Instead, the copolymer resulting from the polymerization of 50:50 molar mixtures of the mono-NF_2 and bis-NF_2-oxetane monomers is an amorphous oil.

Table 5.2 Properties of energetic difluoroamine polymers [4].

Monomer	Monodifluoroamine (A)	Bisdifluoroamine (B)	A/B
Polymer form	Amorphous oil	Solid, m.p. = 158 °C	Amorphous oil
T_g (°C)	−21	130.78	—
Decomposition onset (°C)	191.3	210.0	191.7
Decomposition maximum (°C)	230.7 °C(DSC)	222.2	219.8
Heat flow (J/g)	1851	3031	2409

5.5
Miscellaneous Energetic Fluoropolymers

Energetic difluoroamino-groups have been successfully introduced into polyamides (Nylon 6 or Nylon 12) by exposing the polymers to elemental fluorine [45]. Direct fluorination causes cleavage of the amide C–N bond and concomitant formation of the NF_2 chain-end functionalities. These polymers could also be utilized as energetic ingredients in explosive formulations.

Hydroxy-terminated poly(2-fluoro-2,2-dinitroethyl) polynitroorthocarbonate prepolymers (general structure shown in Figure 5.9), were prepared [46] by reacting bis(2-fluoro-2,2-dinitroethyl)dichloroformal with a diol having the general structure $HOCH_2CXW$-A-$CYZCH_2OH$.

The obtained prepolymer can be cured by diisocynate to form energetic polymer binders. The polymer has a melting range of 60–80 °C, density 1.67 g/ml, and the heat of formation is −603.8 cal/g (1 cal = 4.184 J).

Energetic polyformals are prepared by the reaction of fluorodiols with formaldehyde to form hydroxyl terminated polymers (Scheme 5.3) [47].

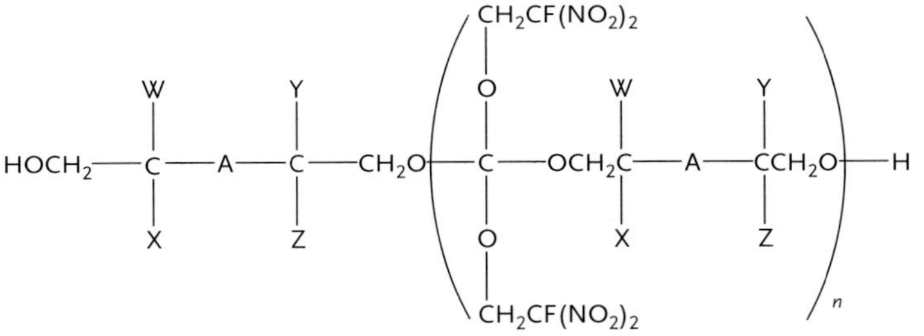

$A = -(CH_2)_{n=1-3}; -(CF2)_{n=1-4}; -(CH_2OCH_2OCH_2)-; -(CH_2OCF_2OCH_2)-; -(CH_2N(NO_2)CH_2)^-$
$W; X; Y; Z = -For-NO_2$

Figure 5.9 General structure of poly(2-fluoro-2,2-dinitroethyl) polynitroorthocarbonate prepolymers [46].

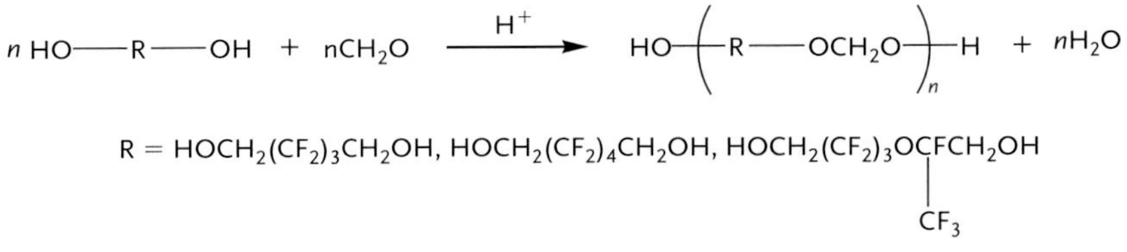

$R = HOCH_2(CF_2)_3CH_2OH, HOCH_2(CF_2)_4CH_2OH, HOCH_2(CF_2)_3OCFCH_2OH$ | CF_3

Scheme 5.3 Preparation of energetic polyformal.

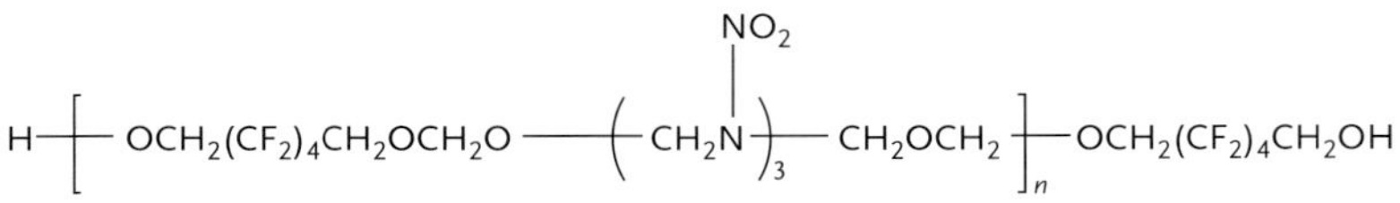

$$\text{H}{\mkern-4mu}\left[\vphantom{\Big|}\right.\text{OCH}_2(\text{CF}_2)_4\text{CH}_2\text{OCH}_2\text{O}\left.\vphantom{\Big|}\right]{-}\left(\text{CH}_2\underset{\underset{\text{NO}_2}{|}}{\text{N}}\right)_3{-}\text{CH}_2\text{OCH}_2\left.\vphantom{\Big|}\right]_n{-}\text{OCH}_2(\text{CF}_2)_4\text{CH}_2\text{OH}$$

Figure 5.10 Structure of the polynitrofluoroformal based energetic polymer [49].

Polymers of M_n values approaching up to 10 000 are obtained in some cases, but are mainly of the order of 2000–4000, including cyclic species. The polymers were resins with low glass transition temperatures in general. The hydroxyl terminated polyformals could be cured using the isocyanate technology to form a cured elastomer. The binder is compatible with an energetic plasticizer, such as bis(2-fluoro-2,2-dinitroethyl) formal, due to the presence of the formal moieties in the polymer backbone [48]. The fluorinated polymers/plasticizers were able to form plastic bonded explosives with HMX with almost the same performance parameters as conventional nitropolymers/plasticizers with HMX [48].

Randomly distributed copolymers of polynitrofluoroformals have been prepared by the polycondensation of nitraminediols and fluorodiols with formaldehyde in sulfolane/boron trifluoride etherate solvent [49] (Figure 5.10).

The hydroxyl end groups of the polymer are derived from the fluorodiol monomer only. The combination of nitraminediol and fluorodiol monomers provides polymers with much lower glass transition temperatures than can be obtained from nitraminediol-based homopolyformals and at the same time provides polymers with significant energy contents [49].

5.6
PBX Formulations with Fluoropolymers

Binders for insensitive explosives do not have to be shock absorbing elastomers. Hence, thermoplastics with glass transitions above the typical operating range, but within the processing range ($80 < T_g < 130\,°C$) of the explosive, are desired as binders for insensitive explosive compositions. Copolymers of fluorinated polymers have found wide application as binders especially for insensitive PBX (polymer bonded explosive) formulations for the past 30 years due to their higher densities (concomitant with the higher fluorine content) and higher thermal stabilities. Fluoropolymer binder PBX formulations are processed by pressing the explosives with binder molding powders. Over the years, several PBX formulations have been developed with insensitive explosives [e.g., TATB, (triaminotrinitrobenzene)] and fluoropolymer binders.

High performing explosive compositions formulated with fluoropolymer binder and HMX and/or TATB have been developed mainly for use with nuclear weapons. These explosives combine the high detonation properties of HMX with the exceptional insensitivity features of TATB.

The performance parameters of selected fluoropolymer based explosive compositions are shown in Table 5.3 [50]. PBX 9502 (95% TATB/5% Kel-F800) and LX-17 (92.5% TATB/7.5% Kel-F800) are widely used as main charge explosives. It has been experimentally proved that Kel-F800 polymer shows the greatest wettability of TATB surfaces and should promote the best adhesion to TATB in PBX formulations [51]. Hence, these PBX compositions could be safely pressed and machined into uniform density parts. They are the most thoroughly studied explosive compositions in use today [52].

Fluorocarbon polymers exhibit excellent thermal stability and have been used in several insensitive booster compositions. PBX 9503 (TATB 80%, HMX 15%, Kel-F800 5%) in which a part of the TATB is replaced with HMX for safety-performance tradeoff, was developed as a booster explosive for PBX 9502. PBXW 7 Type I (TATB 60%, RDX 35%, Teflon 5%) and Type II (TATB 60%, RDX 35%, Viton A 5%) are fluoropolymer binder based booster explosive formulations developed at the Naval Surface Weapons Center, USA [53]. The United Kingdom has developed the BX series of insensitive booster explosive compositions containing RDX/HMX/TATB/Teflon or Kel-F800 [54].

High velocity explosive formulations have been developed with HMX and Viton A (LX-07 and LX-10). These explosives are used in accelerator applications, such as shaped charges, because of their high detonation velocities [55].

Fluoropolymers are poor desensitizers as compared with the polyurethane binders due to the lack of shock absorbing properties of the binder. This feature is

Table 5.3 Detonation properties of selected fluoropolymer based explosives [50, 53].

Explosive	Composition	Density (kg/m^3)	Detonation velocity (m/s)	P_{CJ} (kbar^a)	
LX-07	HMX/Viton A (90 : 10)	1.86	8640	346	High velocity PBX main explosive
LX-10	HMX/Viton A (95 : 5)	1.86	8820	375	High velocity PBX main explosive
LX-17	TATB/Kel-F (92.5 : 7.5)	1.91	7630	250	High velocity insensitive PBX. One of the insensitive IHEs in use
PBX 9502	TATB/Kel-F (95 : 5)	1.94	7700	285	High velocity insensitive PBX. Principal IHE in recent US weapon designs
PBX 9503	TATB/HMX/Kel-F800 (80 : 15 : 5)	1.90	8210	306	Used as a booster explosive in nuclear weapons
PBXW 7 Type I	TATB/RDX/Kel-F800 (65 : 35 : 5)	1.794	7669 (95%TMD and 25.4 mm charge dia.)	—	Insensitive booster explosive

^{a}1 bar = 105 Pa.

exemplified by the fact that the PBXW 7 insensitive booster composition, containing 60% of the highly insensitive TATB, has the ignitability by impact only slightly reduced as compared with the binderless RDX [56]. However, fluoropolymer binders exert influence on one dimensional time to explosion (ODTX) parameters of TATB- or HMX-based compositions, which could be correlated with thermal safety of the explosive. The compositions exhibit longer ODTX in the presence of binders as compared with the pure constituents, consequently improving the thermal stability of the formulation. This feature is attributed to the endothermic decompositions of the binder by reaction with the gaseous decomposition products of HMX and TATB [57]. Because the endothermic decomposition of the binder consumes the intermediate decomposition products of the explosives that are responsible for thermal explosions, more of the explosive should react until the binder is sufficiently decomposed to increase the concentration levels of the decomposition products to the threshold level. The chemical interaction between the Kel-F800 binder and TATB explosive is also reflected in the retardation of the burn rate of the TATB with the addition of the binder [58].

LLM-105 (2,6-diamino-3,5-dinitropyrazine-1-oxide) is an attractive insensitive explosive having higher explosive power and almost similar insensitivity properties to TATB [59]. New experimental fluoropolymer binder (Viton A and Kel-F800) based explosive compositions of LLM-105 have been formulated and characterized [60, 61]. The projected applications of the composition are as new booster formulations having the equivalent performance of HMX. It was demonstrated that mechanically robust pressed booster compositions with superior divergence characteristics (as compared with the binderless ultra-fine TATB composition) could be formulated using the fluoropolymer binders [61].

Hoffman *et al.* [62] studied the mechanical properties of LLM-105 based PBX with different fluoropolymer-based binders. Comparing the profiles of Viton A and Kel-F800 binder based PBXs (Figure 5.11), at $-30\,°C$, which is well below the glass transition temperature of both the binders, the moduli and strengths are similar. On the other hand, at ambient temperature, the modulus and strength of Kel-F800 PBX are higher than those of Viton A PBX, because the glass transition temperature of Kel-F800 is close to ambient and that of Viton A is $-18\,°C$. When the glass transition temperatures of both the polymers have been exceeded ($74\,°C$) the moduli and strengths of the composite PBXs are significantly reduced and become similar.

The strength and modulus of the PBX formulated with Hyflon binder, which is a fully fluorinated semicrystalline copolymer of tetrafluoroethylene (TFE) and 2,2,4-trifluoro-5-trifluoro-methoxy-1,3-dioxide (TTD) with a glass transition temperature of $\sim 150\,°C$, is relatively invariant in the investigated temperature range. Therefore, the glass transition temperature of the binder used in the PBX formulation is a critical parameter influencing its operational temperature.

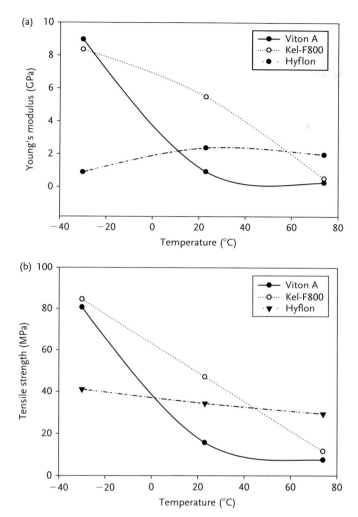

Figure 5.11 Variations of Young's modulus and tensile strength at three different temperatures for PBX compositions based on three different fluoropolymer based binders. Values for the plot taken from Ref. [62].

References

1 Dlott, D.D. (2006) Thinking big (and small) about energetic materials. *Mater. Sci. Tech.*, **22** (4), 463–473.

2 Fawls, C.J., Fields, J.P., and Wagaman, K.L. (2005) Propellants and explosives with fluoro-organic additives to improve energy-release efficiency. US Patent 6,843,868.

3 Dattelbaum, D.M. and Stevens, L.L. (2009) Equations of state of binders and related polymers, in *Static Compression of Energetic Materials* (eds. S.M. Peiris and G.J. Piermarini), Springer.

4 Archibald, T.G., Manser, G.E., and Immoos, J.E. (1995) Difluoroamino oxetanes and polymers formed therefrom to use in energetic formulations. US Patent 5,420,311.

5 Bunn, C.W. and Howells, E.R. (1954) Structures of molecules and crystals of fluoro-carbons. *Nature*, **174** (4429), 549–551.

6 (a) Rae, P.J. and Dattelbaum, D.M. (2004) The properties of PTFE in compression. *Polymer*, **45**, 7615–7625; (b) Brown, E.N., Rae, P.J., Dattelbaum, D.M., Clausen, B., and Brown, D.W. (2008) In-situ measurement of crystalline lattice strains in PTFE. *Exp. Mech.*, **48**, 119–131; (c) Clark, E.S. (1999) Molecular conformations of PTFE: forms II and IV. *Polymer*, **40** (16), 4659–4665; (d) Weeks, J.J., Sanchez, I.C., and Eby, R.K. (1980) Order-disorder transitions in PTFE. *Polymer*, **21** (3), 325–331.

7 Brown, E.N., Trujillo, C.P., Rae, P.J., and Bourne, N. (2007) Soft recovery of PTFE shocked through the crystalline phase II-III transition. *J. Appl. Phys.*, **101** (2), 024916.

8 Eby, R.K., Clark, E.S., Farmer, B.L., Piermarini, G.J., and Block, S. (1990) Crystal structure of PTFE homo- and copolymers in the high pressure phase. *Polymer*, **31**, 2227–2237.

9 Champion, A.R. (1972) Effect of shock compression on electrical resistivity of three polymers. *J. Appl. Phys.*, **43**, 2216.

10 Nagao, H., Matsuda, A., Nakamura, K.G., and Kondo, K. (2003) Nanosecond time resolved Raman spectroscopy on phase transition of PTFE under laser driven shock compression. *Appl. Phys. Lett.*, **83**, 249.

11 Casem, D.T. (2008) Mechanical response of a Al/PTFE composite to uniaxial compression over a range of strain rates and temperatures. Army Research Laboratory, Technical Report Number: ARL-TR-4560, Army Research Laboratory.

12 Yang, Y., Wang, S., Sun, Z., and Dlott, D.D. (2004) Near-infrared laser ablation of PTFE sensitized by nanoenergetic materials. *Appl. Phys. Lett.*, **85** (9), 1493–1495.

13 Parker, L.J., Ladouceur, H.D., and Russell, T.P. (1999) Teflon and Teflon/Al (nanocrystalline) decomposition chemistry at high pressures. *AIP Conf. Proc.*, **505** (1), 941–944.

14 Osborne, D.T. and Pantoya, M.L. (2007) Effect of Al particle size on the thermal degradation of Al/Teflon mixtures. *Combust. Sci. Tech.*, **179**, 1467–1480.

15 (a) Dolgoborodov, A.Y., Makhov, M.N., Kolbanev, I.V., Streletskii, A.N., and Fortov, V.E. (2005) Detonation in an aluminum–Teflon mixture. *JETP Lett.*, **81** (7), 311–314; (b) Dolgoborodov, A.Y., Streletskii, A.N., Makhov, M.N., Kolbanev, I.V., and Fortov, V.E. (2007) Explosive compositions based on the mechanoactivated metal–oxidizer mixtures. *Russian J. Phys. Chem. (B)*, **6**, 606–611; (c) Dolgoborodov, A.Y., Makhov, M.N., Streletskii, A.N., Kolbanev, I.V., Gogulya, M.F., Brazhnikov, M.A., and Fortov, V.E. (2003) Influence of mechanochemical activation on the reactivity of metal-oxidizer mixtures. Proceedings of the 9th International Workshop on Combustion and Propulsion, Paper 30-1-10, Lerici, Italy.

16 Watson, K.W., Pantoya, M.L., and Levitas, V.I. (2008) Fast reactions with nano and micrometer aluminum: a study on oxidation versus fluorination. *Combust. Flame*, **155**, 619–634.

17 Levitas, V.I., Asay, B.W., Son, S.F., and Pantoya, M.L. (2007) Mechanochemical mechanism for fast reaction of metastable intermolecular composites based on dispersion of liquid metal. *J. Appl. Phys.*, **101**, 083524.

18 Hunt, E.M., Malcolm, S., Pantoya, M.L., and Davis, F. (2009) Impact ignition of nano and micron composite energetic materials. *Int. J. Impact Eng.*, **36**, 842–846.

19 Rai, A., Lee, D., Park, K., and Zachariah, M.R. (2004) Importance of phase change of aluminum in oxidation of aluminum nanoparticles. *J. Phys. Chem. B*, **108**, 14793–14795.

20 Committee on Advanced Energetic Materials and Manufacturing Technologies (2004) *Advanced Energetic Materials*, National Research Council,

ISBN 0-309-09160-8 (http://nap.edu/catalog/10918.html). Accessed on 9th October 2008.
21. Grudza, M.E., Jann, D., Forsyth, C., Lacy, W., Hoye, W., and Schaffer, W.E. (2001) Explosive launch studies for reactive material fragments. Proceedings of 4th Joint Classified Bombs/Warheads and Ballistics Symposium, Newport, Rhode Island.
22. Ames, R.G., Garrett, R.K., and Brown, L. (2002) Detonation-like energy release from high speed impacts of PTFE-aluminum projectiles. Proceedings of 5th Joint Classified Bombs/Warheads and Ballistics Symposium, Colorado Springs, Colorado.
23. Schiers, J. (1997) in *Modern Fluoropolymers* (ed. J. Scheirs), Wiley-VCH Verlag GmbH, Weinheim.
24. Drobny, J.G. (2001) *Technology of Fluoropolymers*, CRC Press, Boca Raton, Florida.
25. Dattelbaum, D.M., Robbins, D.L., Sheffield, S.A., Orler, E.B., Gustavsen, R.L., Alcon, R.R., Lloyd, J.M., and Chavez, P.J. (2006) Quasi-static and shock compressive response of fluorinated polymers: Kel-F800, in *Shock Compression of Condensed Matter*, Vol. 845 (eds. M.D. Furnish, M. Elert, T.P. Russell, and C.T. White), American Institute of Physics, pp. 69–72.
26. (a) Piermarini, G. (2009) *In Static Compression of Energetic Materials*, Edited by M. Suhithi Peiris and Gasper. J. Piermarini. Springer; (b) DePiero, S.C. and Hoffman, D.M. (August 2009) Formulation and characterization of LX-17-2 from new FK 800 binder and WA, ATK and BAE TATBs. LLNL-TR-416360.
27. Blumenthal, W.R., Gray, G.T. III, Idar, D.J., Holmes, M.D., Scott, P.D., Cady, C. M., and Cannon, D.D. (2000) Influence of temperature and strain rate on the mechanical behaviour of PBX 9502 and Kel-F800, in *Shock Compression of Condensed Matter*, Vol. 505 (eds. M.D. Furnish, L.C. Chhabildas, and R.S. Hixson), American Institute of Physics, pp. 671–674.
28. Bourne, N.K. and Gray, G.T. III (2005) Dynamic response of binders; teflon, estane and Kel F-800. *J. Appl. Phys.*, **98**, 123503.
29. Veeser, L.R., Gray, G.T., Vorthman, J.E., and Hayes, D.B. (1999) High pressure response of a high purity Iron, In the proceedings of conference on shock compression of matter, American Physical Society, Utah, USA.
30. Hemsing, W.F. (1979) Velocity sensing interferometer modification. *Rev. Sci. Instrum.*, **50**, 73–78.
31. Carter, W.J. and Marsh, S.P. (1995) Hugoniot equation of state of polymers. Los Alamos National Laboratory Report, LA-12006-MS, Los Alamos National Laboratory.
32. Morris, C.E., Fritz, J.N., and Mcqueen, R.G. (1984) The equation of state of PTFE to 80 GPa. *J. Chem. Phys.*, **80**, 5203.
33. Robbins, D.L., Shefield, S.A., and Alcon, R.R. (2004) Magnetic particle velocity measurements on shocked Teflon. Proceedings of the Conference of American Physical Society Topical Group on Shock Compression of Matter, 706, 675–678.
34. Champion, A.R. (1971) Shock compression of teflon from 2.5 to 25 kbar. *Evidence Shock Induced Trans.*, **42** (13), 5546–5550.
35. Baker, B.B. Jr and Kasprzak, D.J. (1993) Thermal degradation of commercial fluoropolymers in air. *Polym. Degrad. Stab.*, **42**, 181–188.
36. (a) Degteva, T.G., Sedova, I.M., and Kuzminskii, A.S. (1963) Thermal degradation of a fluorinated elastomer of the Kel-F type at temperatures above 300 °C- II. *Polym. Sci. USSR*, **4** (5), 1036–1044; (b) Degteva, T.G. and Kuzminskii, A.S. (1964) Oxidative degradation of the fluorine containing elastomer of the Kel-F type in the temperature range of 250–360 °C- I. *Polym. Sci. USSR*, **5** (3), 511–517; (c) Degteva, T.G., Sedova, I.M., and Kuzminskii, A.S. (1964) Mechanism of thermal degradation of an elastomer of the Kel-F type in the temperature range of 250–380 °C- IV. *Polym. Sci. USSR*, **5** (4), 582–589; (d) Burnhanm, A.K. and Weese, R.K. (2005) Kinetics of thermal degradation of explosive binders

Viton-A, estane and Kel-F. *Thermochim. Acta*, **426**, 85–92; (e) Zulfiqar, S., Zulfiqar, M., Rizvi, M., Munir, A., and McNeil, I.C. (1994) Study of the thermal degradation of PCTFE, PVDF, and copolymers of CTFE and VDF. *Polym. Degrad. Stab.*, **43**, 423–430.

37 Mark Hoffman, D. (2003) DMA signatures of Viton A and plastic bonded explosives based on this polymer. *Polym. Eng. Sci.*, **43** (1), 139–156.

38 (a) Taguet, A., Ameduri, B., and Boutevin, B. (2005) Crosslinking of vinylidene containing fluoropolymers. *Adv. Polym. Sci.*, **184** 127–211; (b) Smith, J.F. and Perkins, G.T. (1961) The mechanism of post cure of Viton A fluorocarbon elastomer. *J. Appl. Polym. Sci.*, **5**, 460–467.

39 Moran, A.L., Kane, R.P., and Smith, J.F. (1959) Safe processing curing systems for Viton fluoroelastomers. *J. Chem. Eng. Data*, **4** (3), 276–278.

40 Knight, G.J. and Wright, W.W. (1972) The thermal degradation of hydrofluoro polymers. *J. Appl. Polym. Sci.*, **16**, 683–693.

41 Papazian, H.A. (1972) Prediction of polymer degradation kinetics at moderate temperatures from TGA measurements. *J. Appl. Polym. Sci.*, **16**, 2503–2510.

42 Adolph, H.G. and Trivedi, N.J. (2001) Energetic plasticizers containing 3,3-bis (difluoroamino)-1,5-dinitratopentane and method of preparation. US Patent 6,325,876.

43 Ulas, A., Kuo, K.K., and Gotzmer, C. (2001) Ignition and combustion of boron particles in fluorine environments. *Combust Flame*, **127**, 1935–1957.

44 Ye, C., Gao, H., and Shreeve, J.M. (2007) Synthesis and thermochemical properties of NF_2 containing energetic salts. *J. Fluorine Chem.*, **128**, 1410–1415.

45 Solomun, T., Schimanski, A., Strum, H., and Illenberger, E. (2005) Efficient formation of difluoroamino functionalities by direct fluorination of polyamides. *Macromolecules*, **38**, 4231–4236.

46 Lawrence, G.W. and Gilligan, W.H. (1996) Energetic fluoronitro prepolymer. US Patent 5,574,126.

47 (a) Adolph, H.G., Nock, L.A., Goldwasser, J.M., and Farncomb, R.E. (1991) Nitro- and fluoropolyformals. II. Novel polyformals from α, ω-fluoro and nitro diols. *J. Polym. Sci. A. Polym. Chem.*, **29**, 719–727; (b) Adolph, H.G., Goldwasser, J.M., and Koppes, W.M. (1987) Synthesis of polyformals from nitro and fluoro diols. Substituent and chain length effects. *J. Polym. Sci. A. Polym. Chem.*, **25**, 805–822.

48 Adolph, H.G., Goldwasser, J.M., and Lawrence, W. (1991) Energetic binders for plastic bonded explosives. US Patent 4,988, 397.

49 Adolph, H.G. and Cason-Smith, D.M. (1993) Energetic polymer. US Patent 5,266,675.

50 Yeager, J.D., Dattelbaum, A.M., Orler, E. B., Bahr, D.F., and Dattelbaum, D.M. (2010) Adhesive properties of some fluoropolymer binders with the insensitive explosive TATB. *J. Colloid. Interface. Sci.*, **352**, 535–541

51 Shokry, S.A., Shawki, S.M., and Elmorsi, A.K. (1990) TATB Plastic Bonded Compositions. Proceedings of the 21st International Annual Conference of ICT, Karlsruhe, Germany, 1–16.

52 Gustavsen, R.L., Sheffield, S.A., Alcon, R.R., Forbes, J.W., Tarver, C.M., and Garcia, F. Embedded electromagnetic gauge measurements and modelling of shock initiation in TATB based explosives: PBX 9502 and LX-17. LANL Technical Report: LA-UR-01-3339, Los Alamos National Laboratory.

53 Spear, R.J. and Wolfson, M.G. (1989) Determination of detonation parameters of booster explosives at small charge diameters. DSTO Materials Research Laboratory Report, MRL-TR-89-45, DSTO.

54 Agarwal, J.P. (2005) Some new high energy materials and their formulations for specialized applications. *Propellants Explos. Pyrotech.*, **30** (5), 316–328.

55 Wenzel, A.B. (1987) A review of explosive accelerators for hypervelocity

impact. *Int. J. Impact Eng.*, **5** (1–4), 681–692.
56 Spear, R.J., Nanut, V., Redman, L.J., and Dagley, I.J. (1987) Recommended replacements of tetryl in Australian production fuzes and related ordnance. DSTO Material Research Laboratories Report, MRL-R-1089, DSTO.
57 Tarver, C.M. and Koerner, J.G. (2008) Effect of endothermic binders on times to explosion of HMX- and TATB-based explosives. *J. Energetic Mater.*, **26**(1), 1–28.
58 Son, S.F., Berghout, H.L., Bolme, C.A., Chavez, D.E., Naud, D., and Hiskey, M.A. (2000) Burn rate measurements of HMX, TATB, DHT, DAAF, and BTATz. *Proc. Combust. Inst.*, **28**, 919–924.
59 Tran, T.D., Pagoria, P.F., Hoffman, D.M., Cutting, J.L., Lee, R.S., and Simpson, R.L. (2002) Characterization of LLM-105 as an insensitive high explosive material. Proceedings of 33rd International Annual Conference of ICT, Karlsruhe, Germany.
60 Weese, R.K., Burnham, A.K., Turner, H.C., and Tran, T.D. (2006) Exploring physical chemical and thermal characteristics of a new potentially insensitive high explosive: RX-55-AE-5. Proceedings of 34th Annual Conference of North American Thermal Analysis Society, Bowling Green, KY, USA.
61 Tran, T.D., Pagoria, P.F., Hoffman, D.M., Cunningham, B., Simpson, R.L., Lee, R.S., and Cutting, J.L. (2002) Small scale safety and performance characterization of new plastic bonded explosives containing LLM-105. Proceedings of 12th International Detonation Symposium, San Diego, CA.
62 Hoffman, D.M., Lorenz, K.T., Cunningham, B., and Gagliardi, F. (2008) Formulation and mechanical properties of LLM-105 PBXs. 39th International Annual Conference of ICT, Karlsruhe, Germany.

6
Energetic Plasticizers for High Performance

6.1
Introduction

Plasticizers are critical components in explosive and propellant formulations. The role of plasticizers is to modify the mechanical characteristics of polymer binders by making the polymer chains more flexible. In addition to improving the mechanical properties, plasticizers reduce the mix viscosity to improve processing, modify the oxygen balance, and tailor the burn rate characteristics of energetic material composites.

As with polymer binders, plasticizers can be classified into non-energetic and energetic. Examples of non-energetic plasticizers include acetyl triethyl citrate, diethyl adipate, diethyl sebacate, and diethyl octoate. As non-energetic moieties, these materials significantly dilute the total energy of the formulations. Hence molecules containing energetic groups such as nitro or azido moieties were considered for use as energetic plasticizers. Energetic plasticizers, in addition to improving processability and physical properties, increase the overall energy of the formulation, thus playing a crucial role in augmenting the performance of futuristic propellants. They could also allow the explosive content of the formulation to be reduced, which correspondingly reduces the overall sensitivity.

Energetic plasticizers are liquid materials with molecular weights in the range of 200–3000 g/mol. Low molecular weight plasticizers are more effective in the plasticization effect (reducing the glass transition temperature) but they tend to migrate over time to the surface of the composite. This will cause irregularities in combustion and also lead to changes in the chemical composition of the composite. The migration tendency decreases with increasing molecular weight of the plasticizers. Therefore, energetic azido or nitro polymers with molecular weights range of 500–1000 g/mol have gained importance as energetic plasticizers.

The general requirements of good energetic plasticizers are high heat of formation, high oxygen balance, low glass transition temperature, low ability to migrate, low viscosity, high thermal stability, and low impact sensitivity. These requirements are sometimes contradictory to each other and it is difficult to find a plasticizer having the optimum properties.

Energetic Polymers: Binders and Plasticizers for Enhancing Performance, First Edition.
How Ghee Ang and Sreekumar Pisharath.
© 2012 WILEY-VCH Verlag GmbH & Co. KGaA, Weinheim. Published 2012 by WILEY-VCH Verlag GmbH & Co. KGaA

This chapter provides an overview of the energetic plasticizers based on azido and nitro compounds used in explosive and propellant formulations.

6.2
Energetic Plasticizers Based on Azido Compounds

Energetic plasticizers based on azido groups have the advantages of positive heat of formation and low vapor pressure. They are compatible with modern energetic azido binders such as GAP (glycidyl azide prepolymer). By taking advantage of all these properties, several azido-based compounds have been developed for plasticizer applications.

6.2.1
Azido Acetate Ester Based Plasticizers

Drees et al. [1] synthesized a series of azido acetate ester compounds from ethylene glycol (EGBAA, ethylene glycol bis(azidoacetate) ester), diethylene glycol (DEGBAA, diethylene glycol bis(azidoacetate) ester), trimethylolnitromethane (TMNTA, 1,1,1-Tris(hydroxymethyl)nitromethane tris(azidoacetate) ester), and pentaerythrite (PETKAA, pentaerythritol tetrakis(azidoacetate)). The oxygen atoms in the ester group will help to improve the overall oxygen balance of the compound. The plasticizers exhibited good thermal stability, with glass transition temperatures ranging from $-70\,°C$ for EGBAA to $-35.4\,°C$ for PETKAA. EGBAA showed promising results with the energetic binder Poly(NIMMO) (3-nitratomethyl-methyl oxetane) compared with other commonly used nitrate ester plasticizers. Under 50% plasticizer loading, a rubbery product of Poly(NIMMO) binder is formed with a glass transition temperature of $-66.7\,°C$. However, the viscosity of the azido esters increased with the number of azido groups. The average nitrogen content of these plasticizers was 36% (Figure 6.1).

Qui et al. [2] synthesized a new polyazido ester compound, 1,3-di(azidoacetoxy)-2,2-di(azidomethyl) propane (PEAA) (Figure 6.2) with an enhanced nitrogen content of 47.7%. The glass transition temperature of the compound was $-40.3\,°C$ and was thermally stable up to $218\,°C$. Heat of formation and combustion were not reported for this plasticizer. The calculated value of the oxygen balance was found to be -110. The reported viscosity (2000 mPa s at $20\,°C$) was found to be high for this compound, which will affect the processability.

Pant et al. [3] synthesized a new azido ester compound, namely 1,3-bis(azidoacetoxy)-2-azidoacetoxymethyl-2-ethylpropane (P1) (Figure 6.2) for plasticizer applications. The compound exhibited glass transition temperature of $-47\,°C$, and is stable up to $180\,°C$. The viscosity of the compounds were found to be fairly low (82 mPa s), which is beneficial for the processability. The azido ester exhibited good compatibility with GAP and also showed high heat release rates and insensitivity towards impact and friction.

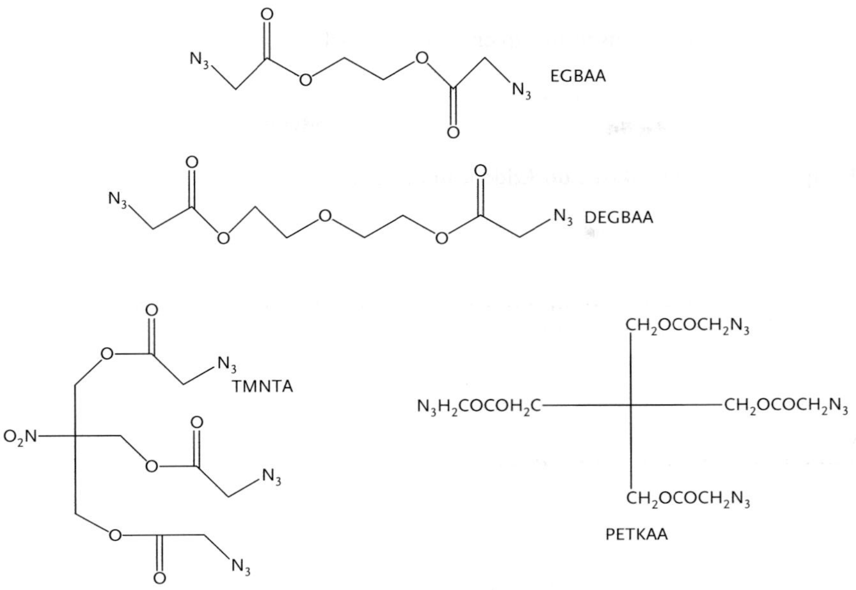

Figure 6.1 Chemical structures of EGBAA, DEGBAA, TMNTA, and PETKAA [1].

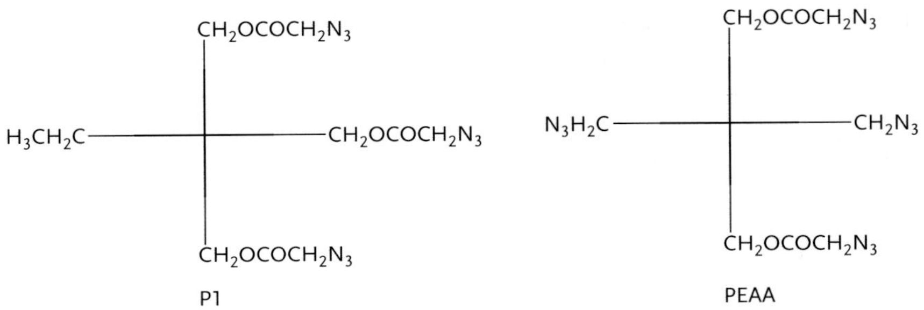

Figure 6.2 Chemical structures of P1 and PEAA.

The physical and thermal properties of azido ester plasticizers are presented in Tables 6.1 and 6.2.

The energetic plasticizer PEAA has the highest heat of formation among the compounds. The glass transition temperatures of all the plasticizers are low enough for propellant applications, the lowest point being −70.8 °C for EGBAA plasticizer. It is noted that the plasticizer containing a nitro-group (TMNTA) has a better oxygen balance of −71.95%.

All the plasticizers are thermally stable up to 200 °C and are insensitive to friction and impact.

Table 6.1 Physical properties of azido ester plasticizers [1, 3].

Compound	Density (g/l)	Glass transition temperature (°C)	Oxygen balance (%)	Heat of combustion	Heat of formation (kJ/mol)	Viscosity at 20 °C (mPa s)
EGBAA	1.34	−70.8	−84.15	3344 kJ/mol	−167.36	23.4
DEGBAA	1.00	−63.3	−99.92	4540 kJ/mol	−328.36	29.2
TMNTA	1.45	−34.1	−71.95	5435 kJ/mol	−232.84	1288
PETKAA	1.39	−35.4	−88.82	7202 kJ/mol	−215.2	2880
P1	–	−47	−110	2129.6 J/g	−157.5	82
PEAA	–	−51.9	−90.8	3096.7 J/g	605.6	205

Table 6.2 Thermal and sensitivity properties of azido ester plasticizers [1, 3].

Compound	DSC onset (°C)	Decomposition peak-TGA (°C)	Friction sensitivity	Impact sensitivity
EGBAA	206.4	218	165 N	5.5 N
DEGBAA	215	212	160 N	>10 N
TMNTA	217.2	207.7	192 N	16 N
PETKAA	221.5	212	360 N	60 N
P1	254.5 (peak)	–	36 Kg	>170 cm
PEAA	245.8 (peak)	–	36 Kg	>170 cm

Agarwal et al. synthesized bis(2-azidoethyl) adipate energetic plasticizer by reacting bis(2-chloroethyl) adipate with sodium azide. The properties of the synthesized plasticizers suggest that the compound could be a potential replacement for the inert plasticizers [4].

6.2.2
Azido Based Oligomeric Plasticizers

An effective approach to preventing the problem of migration of plasticizers is to design oligomeric plasticizers that have some resemblance to polymer binders. Taking this approach into consideration, oligomeric GAP polymers have been developed for energetic plasticizer applications.

Oligomeric GAP polymers are synthesized in a single-step process involving azide displacement of chlorine from epichlorohydrin monomer followed by polymerization without a catalyst [5] (Scheme 6.1). The initiator for the ring opening polymerization of the epichlorohydrin monomer can either be a diol (e.g., ethylene glycol) or a triol (e.g., trimethylolpropane). GAP plasticizers are found to be compatible with GAP binders [6].

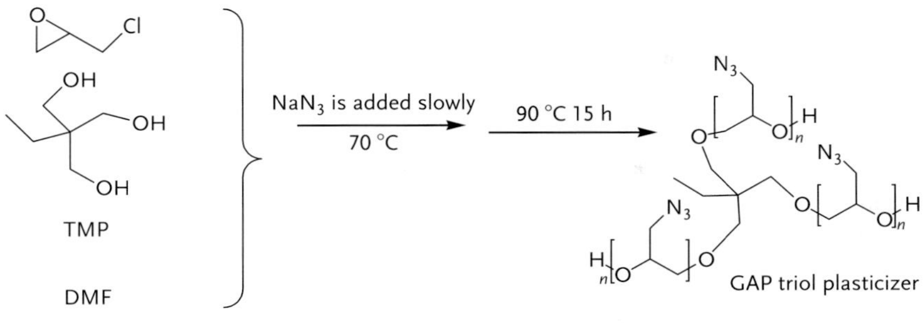

Scheme 6.1 Synthesis of low molecular weight GAP plasticizers.

Ampleman [7] developed the synthesis of an azide terminated glycidyl azide (GAPA) (Scheme 6.2) plasticizer that is more energetic than the hydroxyl terminated GAP polymer.

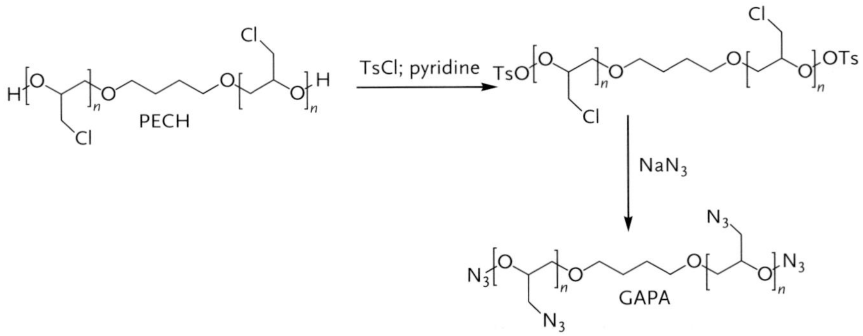

Scheme 6.2 Synthesis of azido terminated GAP plasticizers.

The preparative route of GAPA involves two steps. In the first step, the hydroxyl end groups of the polyepichlorohydrin are converted into tosylate derivatives by reaction with *p*-toluenesulfonyl chloride in anhydrous pyridine. In the second step of the synthesis, the tosylate groups are converted into azido groups by the reaction of sodium azide in N,N-dimethyl formamide.

The properties of GAP triol and GAPA plasticizers are compared in Table 6.3.

GAPA has a higher heat of formation, higher heat of decomposition, and lower glass transition temperature than GAP triol. Both molecules have an oxygen balance of approximately ~−120% [8].

At our Institute, we investigated the plasticization effect of GAP triol and GAPA in cured GAP binders (Figure 6.3).

Free-volume theory forms the theoretical basis of the plasticization action. It proposes that addition of the plasticizer increases the free volume of the plasticized polymer, thus decreasing the glass transition temperature. As shown in

Table 6.3 Properties of GAP triol and GAPA energetic plasticizers.

Property	GAP triol	GAPA
Heat of formation (cal/g)[a] [6]	+280	+550
M_n (g/mol)	440	1032
Density (g/ml)	1.25	1.3
Heat of decomposition (J/g)	1554 (at T_{max} of 239 °C)	2204 (at T_{max} of 240 °C)
Glass transition temperature (°C)	−59.9	−64.4

[a]From Ref. [6]. Rest of the values are from S. Pisharath and H.-G. Ang unpublished results. 1 cal = 4.184 J.

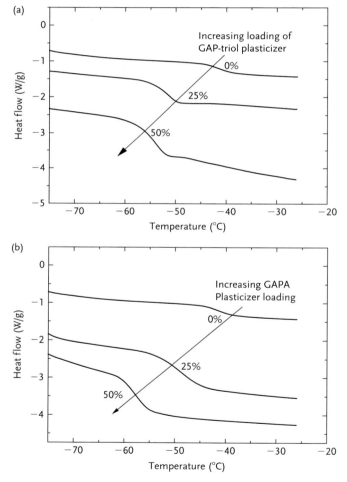

Figure 6.3 DSC thermograms of cured GAP plasticized samples loaded with different amounts of (a) GAP triol plasticizer and (b) GAPA plasticizer. (S. Pisharath and H.G. Ang unpublished results).

Figure 6.3, the glass transition temperatures of the cured GAP samples mixed with GAP triol and GAPA decrease with increasing plasticizer loading. The magnitude of the decrease in the glass transition temperature is higher for GAPA ($\Delta T_g = 16.3\,°C$) as compared with GAP triol ($\Delta T_g = 12.4\,°C$), demonstrating that the plasticization efficiency of the GAPA plasticizer is higher with respect to GAP.

In another study, we compared the loss due to migration of GAP triol plasticizer with that of inert dibutyl phthalate (DBP) plasticizer from cured HTPB samples. Diffusion coefficients at different temperatures were measured by applying Crank's migration model [9] to the data from weight loss measurements.

The magnitude of the diffusion coefficient of the GAP triol oligomeric plasticizer is lower by almost two orders of magnitude as compared with that of the DBP plasticizer (Figure 6.4). Activation energy of diffusion calculated by applying Arrhenius kinetics to the diffusion coefficient–temperature profile demonstrates that the GAP triol oligomeric plasticizer has a higher value of 73 kJ/mol versus 52 kJ/mol for DBP plasticizer. Clearly, azido based oligomeric plasticizers provide advantages of good plasticization efficiency and superior migration resistance.

Pant and *et al.* [10] developed new energetic plasticizers based on dendrimers, which are three dimensional branched polymers with all the bonds emanating from a central core. Owing to their unique macromolecular architecture, dendrimers exhibit functional properties, such as low viscosity, that are advantageous for plasticizer applications. Taking advantage of these positive properties of dendrimers, a second generation of terminally azido functionalized dendritic esters (Figure 6.5) was synthesized in a two-step route from a commercially available dendritic polyol for energetic plasticizer applications [10].

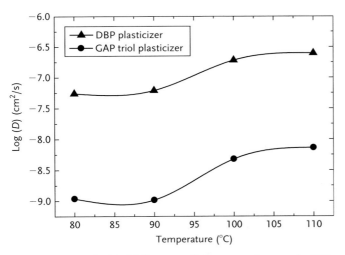

Figure 6.4 Dependence of diffusion coefficient on temperature for DBP and GAP triol plasticizer. (S. Pisharath and H.G. Ang unpublished results).

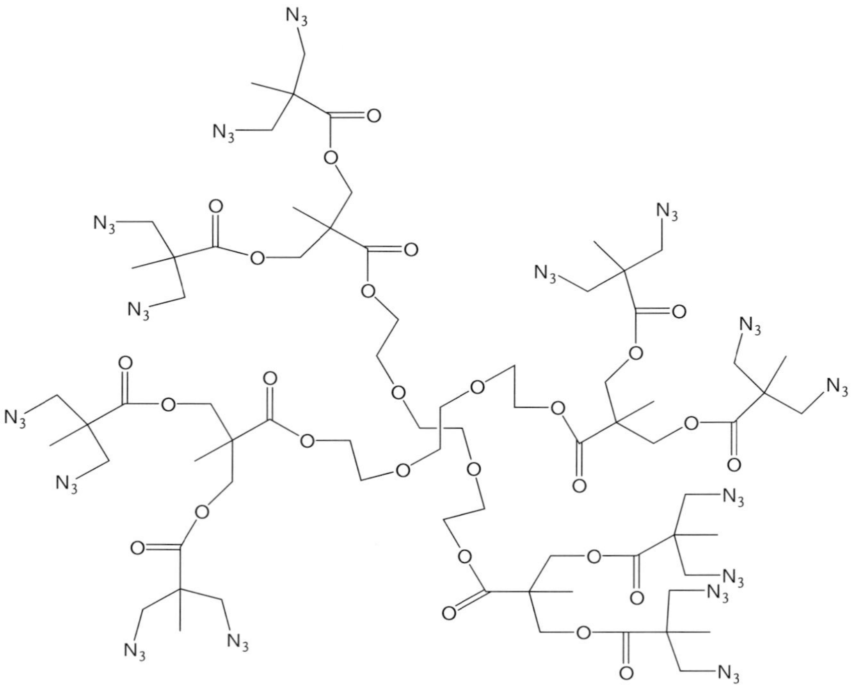

Figure 6.5 Chemical structure of dendritic azido ester plasticizer [10].

The dendritic polyol was tosylated using toluenesulfonyl chloride in pyridine, which was subsequently azidated using sodium azide to produce the azido functionalized dendritic ester.

Thermal studies of the plasticizer by DSC (differential scanning calorimetry) and TGA (thermogravimetric analysis) showed a thermal stability up to 211 °C. DSC revealed that the exothermic heat of decomposition was 1443 J/g at a T_{max} of 248 °C. The glass transition temperature was found to be −35 °C. The compound also exhibited a heat of formation of −141 kJ/mol with an oxygen balance of −130.8%. It is also fairly insensitive to impact (>170 cm) and friction stimuli (>36 kg). The plasticizer shows good compatibility with the azido polymeric binders GAP and BAMO.

6.3
Performance of Propellant Formulations Containing Azido Plasticizers

Theoretical calculations have demonstrated that adding high nitrogen content ingredients to propellants will decrease the exhaust gas molecular weight and flame temperatures without loosing the impetus [11]. Azido and azido ester based plasticizers are being incorporated into experimental advanced gun propellant formulations as replacements of inert plasticizers with the purpose of improving impetus, burn rate and mechanical properties of formulations [12]. Two examples of the application of azido ester and oligomeric energetic plasticizers to advanced gun propellant formulations are given below.

Ghosh et al. studied the effect of azido ester based plasticizers on the ballistic and mechanical properties of triple base gun propellants [12(a)] for tank and field gun munitions. The energy level was found to increase from 1033 to 1040 J/g on replacement of the non-energetic plasticizer dibutyl phthalate (DBP) with 2% of the azido ester plasticizer tris(azido acetoxy methyl) propane (TAAMP) containing three energetic azido groups [12(a)]. The energy is further increased to 1057 J/g when TAAMP is replaced by bis(azido acetoxy) bis(azido methyl) propane (PEAA), a plasticizer containing four azido groups. Because of the better plasticizing effect of azido ester plasticizers, the compression strength of the propellant increases from 257 kg_f/cm^2 for DBP based compositions to 291 kg_f/cm^2 for TAAMP, and 312 kg_f/cm^2 for PEAA compositions, respectively. Thus the azido ester plasticizers are capable of concomitantly enhancing the ballistics and mechanical properties of triple base gun propellants.

Damse et al. studied the performance of GAP plasticizer based RDX propellants using closed vessel firing measurements. The results were compared with the corresponding inert plasticizer, dioctyl phthalate (DOP), based ones and the results are presented in Table 6.4 [12(d)].

GAP based plasticizer propellants were observed to be superior to the DOP-based propellants with respect to the ballistic performance, especially the force constant. The higher force constant is due to the low molecular weight gas products formed, high positive heat of formation, and higher density of GAP compared with DOP. GAP-based propellants were also found to have lower value of pressure exponents, with a higher linear burning rate. These are the promising attributes of the GAP plasticizer based propellants. However, their mechanical properties were lower than those of the DOP-based propellants, but were within the usable limits.

Table 6.4 Comparison of ballistic performances for RDX propellants based on energetic and inert plasticizers [12(d)].

RDX propellant	Force constant (J/g)	Flame temperature (K)	Pressure exponent	Linear burning rate coefficient (cm/s/MPa)
GAP plasticizer based	1275	3530	0.80	0.14
DOP plasticizer based	1200	3210	0.90	0.15

6.4
Nitrate Ester Plasticizers

6.4.1
General Characteristics

Nitrate ester based energetic plasticizers have found wide acceptance due to their ability to enhance the energetic performance of the formulation. Other than their energetic performance, nitrate ester plasticizers provide benefits of improving the rheological properties, preventing crystallization of the binder and improving low temperature mechanical properties of the propellant. Propellants containing nitrate ester plasticizers tend to react less violently during slow cook-off [13].

Nitroglycerin (NG) was the first energetic plasticizer to be developed for commercial explosives. Even though NG is still widely used as an efficient plasticizer in many applications, its high sensitivity towards impact and friction and physiological effects due to exposure prompted the scientific community to look for other less sensitive nitrate ester plasticizers.

Some of the major nitrate esters in use today are trimethylol ethane trinitrate (TMETN), triethylene glycol dinitrate (TEGDN), ethylene glycol dinitrate (EGDN), and butanetriol trinitrate (BTTN). The chemical structures are presented in Figure 6.6.

The energy parameters of important nitrate ester plasticizers are represented graphically in Figure 6.7 and the values are compiled in Table 6.5.

Among the energetic plasticizers, the highest values of specific energy and heat of explosion are recorded for nitrate esters having a positive or small negative oxygen balance (e.g., NG, BTTN). BTTN is often used as a replacement for NG in explosive formulations, because of its improved stability.

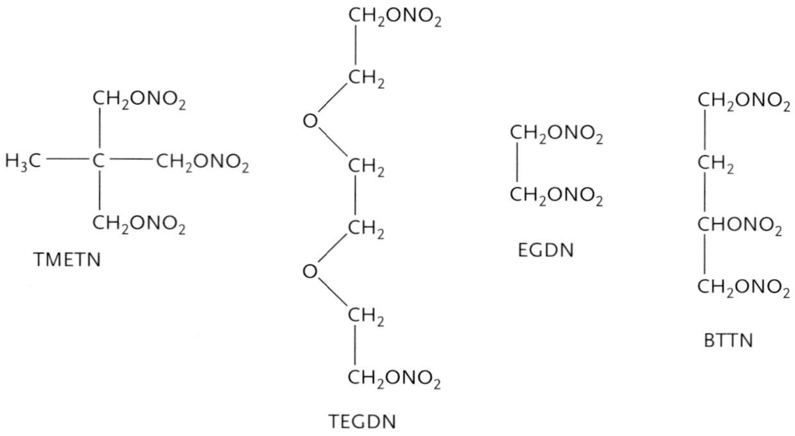

Figure 6.6 Chemical structures of nitrate ester plasticizers.

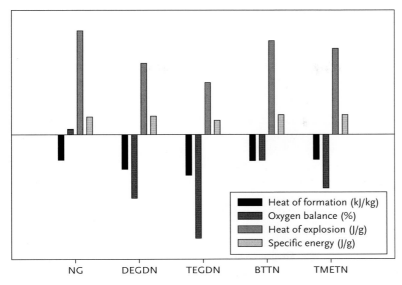

Figure 6.7 Graphical representation of the energy parameters of important nitrate ester plasticizers [14].

Table 6.5 Table of energetic parameters for nitrate ester plasticizers [14].

Plasticizer	Heat of formation (kJ/kg)	Oxygen balance (%)	Heat of explosion (J/g)	Specific energy (J/g)
NG	−1632	+3.5	6671	1125
DEGDN	−2227	−40.8	4566	1178
TEGDN	−2619	−66.6	3317	899
BTTN	−1683	−16.6	6022	1254
TMETN	−1666	−34.5	5530	1260

Thermal decomposition aspects of nitrate esters have been investigated at slow [15] and rapid heating rates [16], which are more characteristic of the combustion. From the low heating rate studies the general agreement was that the decomposition involves an initial first-order rate determining scission of the O–NO$_2$ bond in the molecule, which becomes autocatalytic towards the end of decomposition process. Rapid heating rate studies (2000 °C/s nominal) of nitrate esters were involved with measurement of concentrations of gaseous products evolving out of the decomposition. In the study [16(b)], oxygen balance of the nitrate esters were found to have a strong influence on the characteristics of evolved gases, such as temperature dependence, percentage of hydrocarbons, CO$_2$: CO ratio, and water (Figure 6.8). Both CO$_2$: CO ratio and percentage of H$_2$O increase with the oxygen

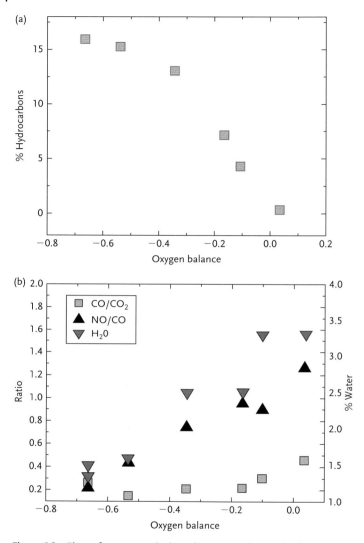

Figure 6.8 Plots of percentage hydrocarbons (a) and several other major product ratios (b) in the evolved gas during thermal decomposition of nitrate ester plasticizers as a function of oxygen balance. Graphs adapted with permission of Wiley from Ref. [16(b)].

balance because of the higher conversion of carbon and hydrogen into respective oxides when more oxygen is present in the nitrate esters. This feature results in the decrease in the percentage of hydrocarbons. As the oxygen balance increases, the amount of NO increases as the number of nitrato groups increase in the molecule and the amount of CO decreases due to extensive oxidation of carbon to give CO_2. These product distribution profiles bear significant influence on flame temperatures and dark zones of propellant compositions using nitrate esters.

The steady-state combustion of the nitrate esters NG, BTTN, and TMETN were modeled by Puduppakkam and Beckstead. [17] using a one-dimensional model. The calculated burn rates were highest for NG and lowest for TMETN. The calculated dark zone lengths and primary flame temperatures were in the order NG>BTTN>TMETN, which is in agreement with the experimental observations.

6.4.2
Performance

In general, incorporation of energetic plasticizers enhances the overall energy and specific impulse of the propellant formulations. The effect of energetic plasticizers in RDX propellant formulations with different binders are illustrated by the specific impulse values (Figure 6.9) [18]. In these formulations, the RDX is 75% by weight and the binder is 25% by weight. The ratio of the plasticizer to binder is 2 : 1 in the mixes. It could be observed that BTTN is more energetic than the TMETN plasticizer. The effect of energetic plasticizer is more pronounced in the HTPB system as compared with the energetic system, as shown by the large difference in the specific impulse between the original and plasticized formulations.

Nitro-ester based energetic plasticizers have also been studied with several other propellant formulations. For the system AP/HMX/GAP/TMETN, the specific impulse of the formulation increases to the range of 2550 N s/kg with the amount of TMETN energetic plasticizer in the binder [19]. The energetic performance of the formulation could be improved further by including the more energetic BTTN into the plasticizer proportion of the TMETN [19]. For the AN/GAP system, the efficiency of BTTN/TMETN plasticizer on improving the burn rate of

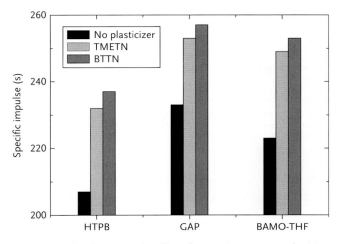

Figure 6.9 Plot illustrating the effect of energetic nitro-ester plasticizers on the specific impulse of RDX propellant mixes formulated with three different binders. Values for the plot are taken from Ref. [18].

formulation could be enhanced by using a new catalyst system based on vanadium/molybdenum oxides along with the formulation [20]. In another application, addition of BTTN to a GAP/RDX mixture helps in balancing the fuel/oxidizer mixture thereby making the combustion of the mixture free from the formation of smoke and char [21]. Thus, as energetic additives, nitrate ester plasticizers are capable of increasing the performance parameters and also modify the burn rate and combustion characteristics of propellant formulations.

Although energetic nitrate ester plasticizers exhibit excellent plasticizing effects with different polymer binders and sufficiently high energetic performances for many propellant/explosive applications, they suffer from the drawback of high shock sensitivity and poor thermal stability. Most of the nitrate esters are classified as HD 1.1 explosives. For example, the GAP/TMETN/RDX composite is more impact sensitive (drop weight impact = 24 cm) than the HTPB/RDX composite (drop weight impact = 52 cm) due to the addition of the sensitive nitrate ester plasticizers [18].

Consequently, new nitro-based plasticizers with high thermal stability were devised to improve the safety characteristics of the formulations. A novel plasticizer 2,2-dinitro-1,3-propanediol diformate (ADDF) was developed [22]. As illustrated in Figure 6.10, the ADDF molecule contains essentially formate ester bonds and is free from nitrate ester bonds, which are more vulnerable to shock and thermal stimuli.

ADDF is prepared by the esterification of 2,2-dinitro-1,3-propanediol. It possesses heat of formation of -2744 kJ/kg, density of 1.47 g/cm^3, and DSC onset of decomposition of 290 °C. It is worth noting that the decomposition temperature of ADDF is higher than that of TEGDN (200 °C), which makes it a safer alternative than the conventional nitrate esters as plasticizers. Melt cast energetic materials could be formulated by using ADDF plasticizers in conjunction with meltable energetic materials, such as TNT or TNAZ (1,3,3-trinitroazetidine) [22].

6.5
Nitrate Ester Oligomers as Energetic Plasticizers

Plasticizers possess an inverse correlation with molecular weight and mobility. The smaller the molecule the quicker it will exude through the binder matrix. Because

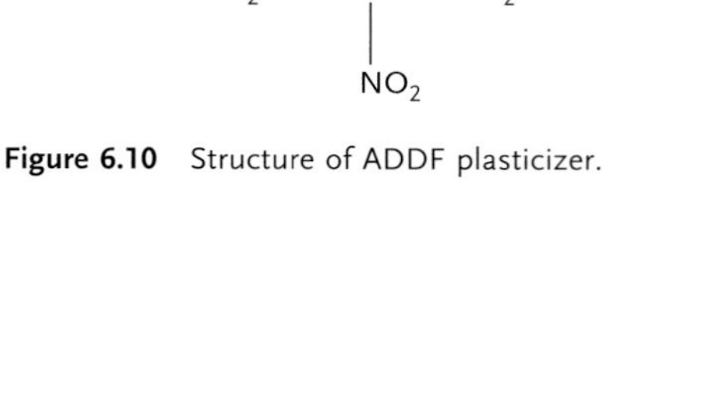

Figure 6.10 Structure of ADDF plasticizer.

of these reasons, polymers have a distinct advantage as plasticizers over small molecules as plasticizers. As in the case of azido polymers, the oligomers of nitropolymers Poly(NIMMO) (3-nitratomethyl-methyl oxetane) and PGN (glycidyl nitrate) have been reported to be effective plasticizers [6, 23]. As exemplified in Scheme 6.3 for PGN, these oligomers have been further nitrated to nitrato-terminated oligomers in order to prevent their cross-linking with the binder system and resultant loss in the plasticization effect [6, 23]. It is expected that nitration improves the oxygen balance and overall energy of the plasticizer system.

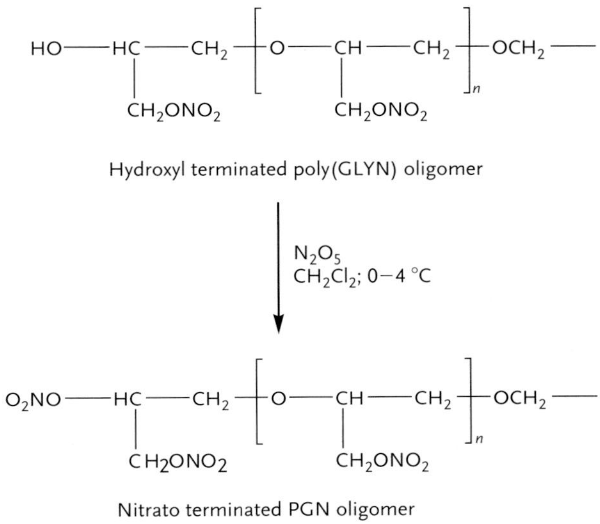

Scheme 6.3 Synthetic of nitrato terminated PGN oligomer

PGN has a number of advantages over traditional nitrate ester plasticizers including low volatility, low glass transition temperature, excellent miscibility with the binder, and decreased plasticizer mobility [24].

Plasticizer migration is a common problem encountered in energetic material formulations. Provotas [25] compared the diffusion coefficients of energetic GN oligomer and K-10 obtained by isothermal thermogravimetric analysis (Figure 6.11). K-10 plasticizer is a 65 : 35 mixture of dinitroethylbenzene and trinitroethylbenzene.

It is observed that the migration rates are inversely proportional to the molecular weight of the plasticizer. Therefore, the migration rate of the PGN oligomer is lower than that of the K-10 plasticizer. Compared with the K-10 plasticizer, the slower migration rate of PGN will help to retain the plasticizer in the formulation under prolonged ageing and prevent the deterioration of its mechanical properties.

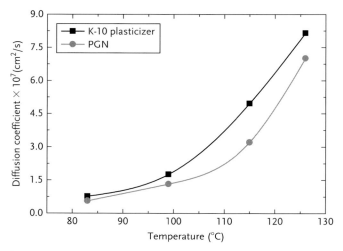

Figure 6.11 Comparison of diffusion coefficients for GN oligomer and K-10 plasticizer. Values for the plot are taken with permission of Taylor & Francis Inc. from Ref. [25].

6.6
Miscellaneous Plasticizers Based on Nitro-Groups

6.6.1
Polynitro-Aliphatic Plasticizers

Research on polynitro-aliphatic compounds commenced in the USA in the 1940s with the objective of developing energetic plasticizers taking advantage of their higher density and higher oxygen content. BDNPA/F is a low cost energetic plasticizer that was developed out of this research initiative in the 1950s for the Polaris space shuttle program. This plasticizer has found extensive application in explosive/propellant formulations [26, 27]. It is a 1 : 1 eutectic mixture of bis(2,2-dinitropropyl) formal and bis(2,2-dinitropropyl) acetal (Figure 6.12).

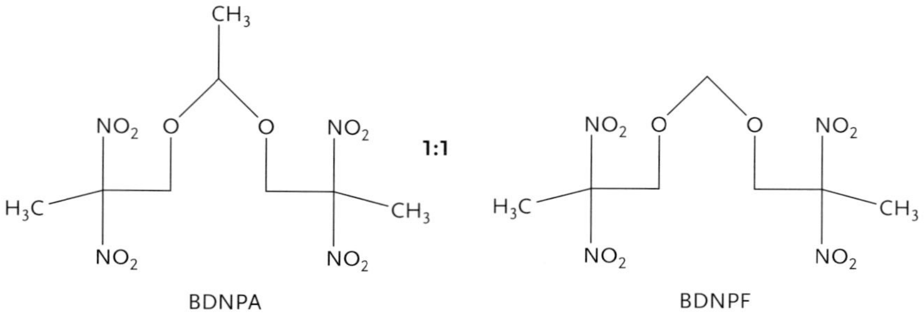

Figure 6.12 Structure of BDNPA/F plasticizer.

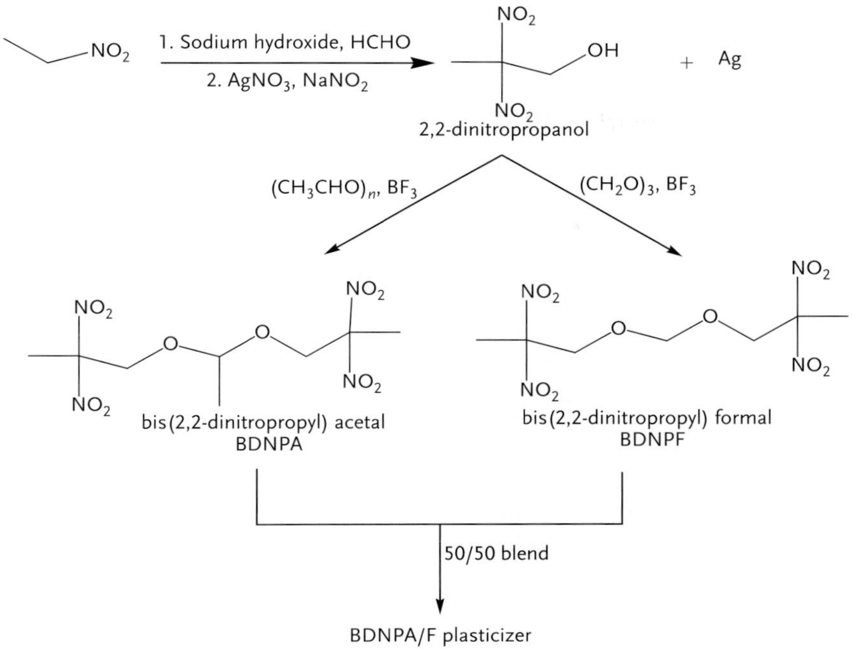

6.6 Miscellaneous Plasticizers Based on Nitro-Groups

Scheme 6.4 Synthesis of BDNPA/F Plasticizer

BDNPA/F plasticizer was synthesized by US Navy and Aerojet in the 1960s by the oxidative nitration of nitroethane to 2,2-dinitropropanol followed by the reaction with acetaldehyde and formaldehyde as shown in Scheme 6.4 [26, 27].

BDNPA/F is currently manufactured using an environment friendly synthetic methodology developed by Thiokol. Thiokol is also working on an electrochemical synthetic route to oxidize nitroethane into dinitropropanol by completely eliminating the environmentally polluting inorganic salts in the process route [28].

Properties of BDNPA/F are presented in Table 6.6.

BDNPA/F is an extensively used energetic plasticizer in LOVA propellant formulations and IM explosive formulations due to its insensitivity attributes [30].

Table 6.6 Properties of BDNPA/F [29].

Melting-point (°C)	−15
Oxygen balance (%)	−57.6
Density (g/ml)	1.39
Molecular weight (g/mol)	319.2
Glass transition temperature (°C)	−65.2
Enthalpy of formation (kJ/kg)	−1940

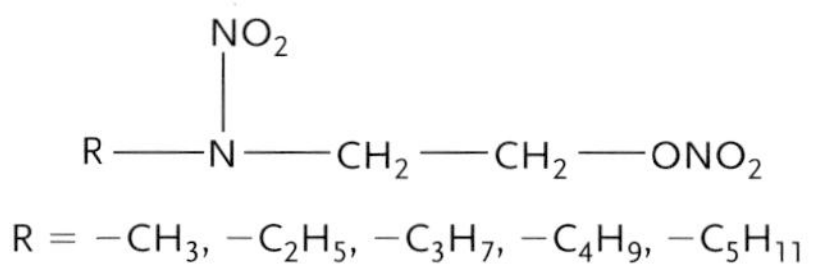

$$\text{R}-\underset{|}{\text{N}}-\text{CH}_2-\text{CH}_2-\text{ONO}_2$$
$$\overset{\text{NO}_2}{}$$

R = $-CH_3, -C_2H_5, -C_3H_7, -C_4H_9, -C_5H_{11}$

Figure 6.13 General structure of NENA plasticizer.

It is used in 13 out of the 24 recognized explosive formulations developed by Picatinny Arsenal (PAX series) [26(a)]. The most prominent among them are PAX-2A (85%HMX, 9%A/F, 6%CAB) and the blast explosive PAX-3 (84% (HMX + aluminum), 9.5% A/F, 6.5%CAB).

The problem with BDNPA/F plasticizer is the lower thermal and chemical stability of the acetal group compared with that of the formal group, thereby deteriorating the overall thermal stability of the plasticizer. To curtail this problem, Cho et al. at the Agency for Defense Development, Korea, developed a new plasticizer comprising a eutectic mixture of bis(2,2-dinitropropyl) formal (BDNPF) and bis(2,2-dinitropropyl) diformal (BDNPDF) [31]. The BDNPDF component is used to prevent the BDNPF from crystallizing out. The energetic properties of this plasticizer have not yet been reported.

6.6.2
Nitratoethylnitramine (NENA) Plasticizers

Nitratoethylnitramines (NENA) are another class of high performance energetic plasticizers used in energetic material formulations particularly in nitrocellulosic systems. Important NENA compounds include the "N-alkyl-N-nitratoethylnitramine" compounds, such as the butyl, ethyl, and methyl derivatives. NENA compounds contain both nitrate ester and nitramine functionalities, having the general structure shown in Figure 6.13.

NENA plasticizers provide the highest impetus at any flame temperature than any of the other energetic ingredients [32]. NENAs are manufactured by the nitration of commercially available alkyl ethanolamines (Scheme 6.5).

$$\text{RNHCH}_2\text{CH}_2\text{OH} \xrightarrow{98\% \text{ HNO}_3} \text{RN(NO}_2)\text{CH}_2\text{CH}_2\text{ONO}_2$$

Scheme 6.5 General scheme for synthesis of NENA Plasticizers

Properties of important NENA plasticizers are presented in Table 6.7.

The heat of formation of NENA plasticizers is higher than that of BDNPA/F and other nitrate ester plasticizers. Hence, NENA plasticizers are possible replacements for the less energetic BDNPF/A plasticizers.

Cartwright studied the volatile nature of NENA plasticizers from a nitrocellulose based propellant matrix using thermogravimetric analysis (TGA) [34]. The results

Table 6.7 Properties of NENA plasticizers [33].

Properties	Me-NENA	Et-NENA	Pr-NENA	Bu-NENA	Pe-NENA
Molecular weight	165.1	179.1	193.2	207.2	221.1
Density (g/ml)	1.51	1.32	1.264	1.211	1.178
O_2-balance (%)	−43.6	−67	−87	−104	−119.1
Heat of formation (kJ/mol)	–	−142	−154	−167	–

were compared with those for other plasticizers such as DEGDN (diethylene glycol dinitrate) and NG. In general, the NENA plasticizers had lower vaporization rates than DEGDN and NG, making it a prospective energetic plasticizer. The plasticizer diffusivity and activation energy of plasticizer vaporization strongly depend on the molecular structure. The differences in molecular structure contribute to the entropic component of the activation energy. Among the NENA plasticizers, butyl NENA shows anomalously low diffusivity, while it was very high for the nitrate ester NG. It is anticipated that the entropic effect due to the compactness of the NG molecule is responsible for its easy vaporization and high volatility. On the other hand, the vaporization of butyl NENA is inhibited by its elongated and asymmetrical structure. The percentage nitrogen content of the Nitrocellulose (NC) matrix also influences the activation energy of vaporization of NENA plasticizers. The activation energy of vaporization of methyl NENA decreases with respect to that of ethyl NENA on moving from an NC of 13.25% N to 12.6% N. It is believed that the more polar and hydrogen bonded structure of NC containing 12.6% N facilitates the easy diffusion of methyl NENA but hinders the diffusion of less polar plasticizers. For butyl NENA, both activation energy and diffusion relative to ethyl NENA increases on going from 13.25%N–NC to 12.6% N–NC.

Introduction of butyl NENA plasticizer to propellant formulations has resulted in concomitant improvement of mechanical properties and energetics [35]. In single-base gun propellant formulations, introduction of butyl NENA provides significant gain in force constant and flame temperature without affecting the linear burn-rate coefficient. The substitution of inert diethyl phthalate with butyl NENA in double-base propellants results in appreciable gain in energy (force constant and flame temperature) as well as the compressive strength. Introduction of butyl NENA in triple-base propellants lead to an appreciable drop in the sensitivity with some improvements in the mechanical properties.

References

1 Drees, D., Loffel, D., Messmer, A., and Schmid, K. (1999) Synthesis and characterization of azido plasticizer. *Propellants Explos. Pyrotech.*, **24**, 159.

2 Qiu, S., Fan, H., Gao, C., Zheng, X., and Gan, X. (2006) An azido ester plasticizer, 1, 3-di (azidoacetoxy) -2.2-di (azidomethyl) propane (PEAA). Synthesis, characterization and thermal

properties. *Propellants Explos. Pyrotech.*, **31** (3), 205.

3 Pant, C.S., Wagh, R.M., Nair, J.K., Gore, G.M., and Venugopalan, S. (2006) Synthesis and characterization of two potential energetic azido esters. *Propellants Explos. Pyrotech.*, **31** (6), 477.

4 Agarwal, J.P., Bhongle, R.K., David, F.M., and Nair, J.K. (1993) Bis (2-azido ethyl) adipate plasticizer: synthesis and characterization. *J. Energ. Mater.*, **11** (1), 67–83.

5 Ahad, E. (1990) Direct conversion of epichlorohydrin to glycidyl azide polymer. US Patent 4,891,438.

6 Provatas, A. (2000) DSTO - Energetic polymers and plasticizers for explosive formulations. Aeronautical and Maritime Research Laboratory Technical Report Number: DSTO-TR-0966, DSTO, Maritime Research Laboratory.

7 Ampleman, G. (1992) Synthesis of diazido terminated energetic plasticizer. US Patent 5,124,463.

8 Bohn, M.A. (1999) Determination of kinetic data of the thermal decomposition of energetic plasticizers and binders by adiabatic self heating. *Thermochim. Acta*, 121–139.

9 Crank, J. (1956) *Mathematics of Diffusion*, Clarendon Press, Oxford.

10 Pant, C.S., Wagh, R.M., Nair, J.K., and Mukundan, T. (2006) Dendtritic azido ester: a potential energetic additive for high energy material formulations. *J. Energ. Mater.*, **24**, 333–339.

11 (a) Zentner, B.A. and Reed, R. (1992) Calculated performance of gun propellant compositions containing high nitrogen ingredients. 5th International Gun Propellant and Propulsion Symposium. New Jersey; (b) Volk, F. and Bathelt, H. (1997) Influence of energetic materials on the energy-output of gun propellants. *Propellants Explos. Pyrotech.*, **22**, 120.

12 (a) Ghosh, K., Pant, C.S., Sanghavi, R., Adhav, S., and Singh, A. (2009) Studies on triple base gun propellants based on two energetic azido esters. *J. Energ. Mater.*, **27**, 40–50; (b) Sanghavi, R.R., Asthana, S.N., Karir, J.S., and Singh, H. (2001) Studies on thermoplastic elastomers based RDX-propellant compositions. *J. Energ. Mater.*, **19**, 79–95; (c) Sanghavi, R.R., Kamale, P.J., Shaikh, M.A.R., Shelar, S.D., Sunilkumar, K., and Singh, A. (2007) HMX based enhanced energy gun propellant. *J. Hazardous Mater.*, **143**, 532–534; (d) Damse, R.S., Singh, A., and Singh, H. (2007) High energy propellants for advanced gun ammunition based on RDX, GAP and TAGN compositions. *Propellants Explos. Pyrotech.*, **32** (1), 52.

13 Wallace, I.A. II, and Oyler, J. (2003) Nitrate ester plasticized energetic compositions, method of making and rocket motor assemblies containing the same. US Patent 6,632,378.

14 Volk, F. and Bathelt, H. (2002) Performance parameters of explosives: equilibrium and non-equilibrium reactions. *Propellants Explos. Pyrotech.*, **27**, 136.

15 (a) Zeman, S. (1997) Kinetic compensation effect and thermolysis mechanisms of organic polynitroso and polynitro compounds. *Thermochim. Acta*, **290**, 199; (b) Makashir, P.S. and Kurian, E.M. (1999) Spectroscopic and thermal studies on pentaerythritol tetranitrate. *Propellants Explos. Pyrotech.*, **24**, 268; (c) Zeman, S. (1992) New dependence of activation energies of nitroesters thermolysis and possibility of its application. *Propellants Explos. Pyrotech.*, **17**, 17.

16 (a) Chen, J.K. and Brill, T.B. (1991) Thermal decomposition of energetic materials 50. Kinetics and mechanism of nitrate ester polymers at high heat rates by smatch/FTIR spectroscopy. *Combust. Flame*, **85**, 479; (b) Roos, B.D. and Brill, T.B. (2002) Thermal decomposition of energetic materials 82. Correlations of gaseous products with the composition of aliphatic esters. *Combust. Flame*, **128**, 181.

17 Puduppakkam, K.V. and Beckstead, M.W. (2005) Combustion modeling of nitrate esters with detailed kinetics. 41st AIAA Joint Propulsion Conference and Exhibit, Tucson, Arizona.

18 Yee, R.Y. and Martin, E.C. (1984) Effects of surface interactions and mechanical properties of plastic bonded explosives on explosive sensitivity. Part 2: Model formulation. Technical Report of Naval Weapons Center No: TP 6619, Naval Weapons Center, China Lake, CA.

19 Menke, K. and Eisele, S. (1997) Rocket propellants with reduced smoke and high burning rate. *Propellants Explos. Pyrotech.*, **22**, 112.

20 Menke, K., Böhnlein-Mauß, J., and Schubert, H. (1996) Characteristic properties of AN/GAP propellants. *Propellants Explos. Pyrotech.*, **21**, 139.

21 Roos, B.D. and Brill, T.B. (2001) Thermal decomposition of energetic materials 81. Flash pyrolysis of GAP/RDX/BTTN propellant compositions. *Propellants Explos. Pyrotech.*, **26**, 213.

22 Highsmith, T.K., Doll, D.W., and Cannizzo, L.F. (2002) Energetic plasticizer and explosive propellant composition containing same. US Patent 6,425,966.

23 Willer, R., Stern, A.G., and Day, R.S. (1995) Polyglycidyl nitrate plasticizers. US Patent 5,380,777.

24 Agarwal, J.P. and Hodgson, R.D. (2007) *Organic Chemistry of Explosives*, John Wiley & Sons, Ltd, Chichester.

25 (a) Provatas, A. (2003) Energetic plasticizer migration studies. *J. Energ. Mater.*, **21**, 237–245; (b) Provatas, A. (2003) DSTO-Aeronautical and Maritime Research Laboratory Technical Report Number: DSTO-TR-1422, DSTO, Maritime Research Laboratory.

26 (a) Rindone, R., Geiss, D.A., and Miyoshi, H. (2000) BDNPA/BDNPF shows long term stability. Proceedings of Insensitive Munitions and Energetic Materials Technology Symposium, USA; (b) Adolph, H.G. (1992) BDNPF/A. US Patent 4,997,499.

27 Hamilton, R.S. and Wardle, R.B. (1995) Synthesis of bis(2,2 dinitropropyl) formal. US Patent 5,449,835.

28 Lusk, S.K. (2004) Electrochemical Oxidation of Alkylnitro Compounds PP-1345. A Technical Report Submitted by ATK Thikol Inc. to Strategic Environmental Research and Development Program Office.

29 Wingborg, N. and Eldsater, C. (2002) 2,2-Dinitro-1,3-bis-nitrooxy-propane (NPN): a new energetic plasticizer. *Propellants Explos. Pyrotech.*, **27**, 319.

30 Alexander, B., Carrillo, A., and Yim, K.B. (2009) Novel manufacturing process development and evaluation of high blast explosive with BDNPA/F and R8002 plasticizers. Proceedings of Insensitive Munitions and Energetic Materials Technology Symposium, USA.

31 Cho, J.R., Kim, J.S., Lee, K.D., and Park, B.S. (2003) Energetic plasticizer comprising bis (2,2 dinitropropyl) formal and bis (2,2 dinitropropyl) diformal and preparation method thereof. US Patent 6,592,692.

32 Agarwal, J.P. (2010) *High Energy Materials. Propellants, Explosives and Pyrotechnics*. Wiley-VCH, Weinheim.

33 (a) Simpson., R.L. (1994) NENAs. New plasticizers. NIMIC-S-275-94. NATO, Brussels. Belgium; (b) Badgujar, D.M., Talawar, M.B., Asthana, S.N., and Mahulikar, P.P. (2008) Advances in science and technology of modern energetic materials: an overview. *J. Hazard. Mat.*, **151**, 289–305.

34 Cartwright, R.V. (1995) Volatality of NENA and other energetic plasticizers determined by thermogravimetric analysis. *Propellants Explos. Pyrotech.*, **20** (2), 51–57.

35 (a) Rao, K.P.C., Sidker, A.K., Kulkarni, M.A., Bhalerao, M.M., and Gandhe, B.R. (2004) Studies of n-Bu-NENA: synthesis, characterization and propellant evaluations. *Propellants Explos. Pyrotech.*, **29** (2), 93–98; (b) Chakraborthy, T.K., Raha, K.C., Omprakash, B., and Singh, A. (2004) A study of gun propellants based on Bu-NENA. *J. Energ. Mater.*, **22** (1), 41–53; (c) Damse, R.S. and Singh, A. (2008) Evaluation of energetic plasticizers for solid gun propellant. *Defense Sci. J.*, **58** (1), 86–93.

7
Application of Computational Techniques to Energetic Polymers and Formulations

7.1
Introduction

Accurate prediction of energetic materials behavior from the first principles requires the characterization of a wide range of phenomenon with disparate temporal and spatial scales from electrons and atoms to devices. A multiscale, multiphysics approach involving a combination of theories and computational techniques is required to capture the entire phenomenon [1, 2]. The powerful and widely used computational techniques for materials modeling include: (i) first principles quantum mechanics (QM), (ii) large scale molecular dynamics (MD) simulations, (iii) mesoscale modeling, and (iv) micro- and macro-scale modeling.

As illustrated in Figure 7.1, powerful computational techniques capable of modelling a wide range of physical/chemical processes with different spatial and temporal scales have been developed over time. The computational techniques extend from the atomic scale ab initio QM level to the macro scale finite element methods. The developed models permit safety and performance screening of large numbers of candidate energetic materials and composites prior to large scale production. This chapter provides an overview of computational techniques employed and also illustrates the application of these techniques to energetic binders and its composites.

7.2
Overview of Computational Techniques

Quantum mechanics methods provide an accurate, although approximate, description of the chemical bond in a very large number of systems. They offer the possibility of analysis and prediction of stability of molecules that are yet to be synthesized, and the chemical species (transition and excited states) which are difficult to isolate. Even though QM methods are particularly suitable for the investigation of energetic materials [3], they are extremely computer intensive. The system size

Energetic Polymers: Binders and Plasticizers for Enhancing Performance, First Edition.
How Ghee Ang and Sreekumar Pisharath.
© 2012 WILEY-VCH Verlag GmbH & Co. KGaA, Weinheim. Published 2012 by WILEY-VCH Verlag GmbH & Co. KGaA

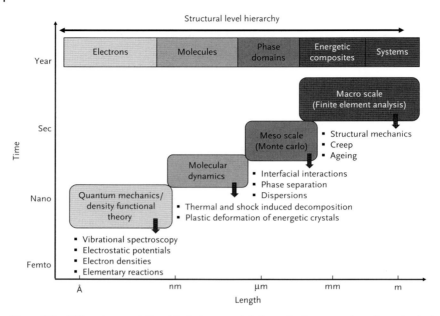

Figure 7.1 Different computational techniques and their application to study various physical processes with characteristic times and distances.

is limited to a maximum of 100 atoms. This has so far precluded the application of QM methods to the study of very large and/or disordered systems.

Molecular dynamics (MD) techniques [4] are capable of dealing with systems of thousands and even millions of atoms and predict static and dynamic behaviors of materials at the molecular level. In MD simulations, molecules move under the action of force fields, which are additive and symmetric and are derived from intermolecular potentials. The dynamic evolution of the system is governed by classical Newtonian mechanics. The empirical force fields are used to predict equilibrium and non-equilibrium properties of condensed systems. MD simulations provide information on atomic positions and velocities at the nanoscale level, from which macroscopic properties (e.g., pressure, energy, and heat capacities) can be derived with the help of statistical mechanics. This approach, while appropriate for systems such as the rare gases, fails for covalent and/or metallic systems. Car and Parinello [5] developed a hybrid computational technique of MD and QM methods (quantum molecular dynamics), which is capable of extending the range of both of the concepts and offers a convenient tool for studying finite temperature effects and dynamical properties.

Despite the widespread use of MD simulations and substantial progress in the related computational methods, the size and timescales of the MD approaches have been practically limited to simulation times and system sizes of less than 100 ns and 10 nm. However, morphological features observed in experiments involve much larger spatial and/or temporal scales. Hence, as illustrated in Figure 7.2, there is a multiscale modeling challenge for validating the molecular models with experimental results. This is due to the lack of rigorous mathematical and computational frameworks providing a direct link of the atomistic scales to complex mesoscopic/macroscopic phenomenon [2].

Coarse-grain (CG) modeling, which represents a system by a reduced number of degrees of freedom, has been proposed as a feasible way to extend the scope of molecular modeling and bridge it with experimental techniques. In CG models, materials are represented as simpler bead-spring models rather than as individual atoms as in MD simulations (Figure 7.3) [6].

Owing to the reduction in the degrees of freedom (typically of the order of 10 : 1) and elimination of fine interaction details, the simulation of (CG) systems requires fewer resources and proceeds faster than that for the same system in an all-atom representation [7]. As a result, an increase of orders of magnitude in the simulated time and length scales can be achieved, by retaining significant atomic scale information.

The integration between the CG models to the details of atomistic simulation is achieved by adopting a multiscale approach using a combination of MD and mesoscale modeling techniques. The important mesoscale simulation methods

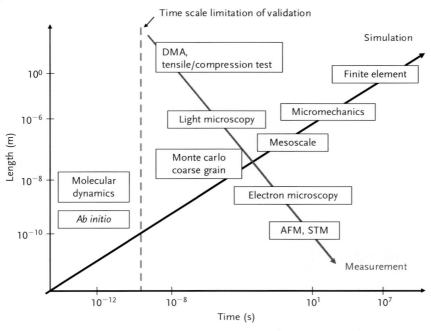

Figure 7.2 Schematic describing the multiscale modeling challenge for probing phenomena in different spatial and temporal scales [2]. Reprinted with permission from Elsevier.

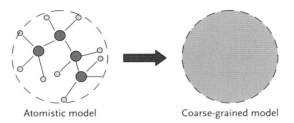

Atomistic model   Coarse-grained model

Figure 7.3 Schematic representation of coarse-grained model.

are: field based dynamic mean field theory (Mesodyn) and particle based dissipative particle dynamics (DPD) [8]. The input parameters for mesoscale simulations (the bead structure and length, the Flory–Huggins χ parameter, and the bead-self diffusion coefficient) are obtained from the MD simulations [6(a)]. In recent years, efforts have been made to bridge the gap between coarse grain and fully atomistic models by using efficient mapping and reverse-mapping techniques [9].

The molecular structure description of the behavior of materials should be homogenized to a continuum level, in order to understand their macroscopic properties. In the continuum level description, the discrete nature of the atomistic and molecular structure is ignored and the material is treated as being continuously distributed throughout the volume, having an average density (Figure 7.4). In the case of composite materials, the chemical interaction between the constituent phases is not explicitly included [10].

The continuum method relates the deformation of a continuous medium in terms of internal stress and strain to external forces acting on it. Computational approaches for the modeling of a continuum description of materials can be classified as either analytical or numerical types. Numerical continuum modeling techniques solve the continuum equations using finite-element or finite-difference techniques. A prominent example is the finite-element analysis (FEA), which uses geometries, volume fractions, and other properties (density, etc.) of the constituent phases for the numerical computation of the bulk properties of the composite

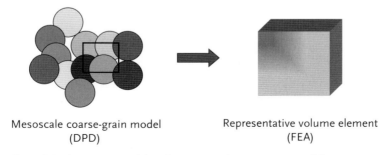

Mesoscale coarse-grain model   Representative volume element
(DPD)   (FEA)

Figure 7.4 Continuum modeling from mesoscale coarse-grain model.

materials. FEA involves the discretization of the representative volume element (RVE) of a continuous medium into simple sub-domains, for which the elastic solutions will provide the desired stress and strain values [11]. Boundary element (BE) is another type of numerical continuum-based method, which only requires elements along the boundaries of the RVE for evaluation of the stress and strain fields [12]. This feature renders it computationally less exhaustive than the FE approach. In addition to the numerical approaches, a multitude of analytical continuum based approaches have evolved based on a broad variety of model approximations, which have been employed for estimating the lower and upper bounds of the mechanical properties of polymer composites [10, 13].

Several computational approaches have evolved over the years for linking the continuum models with atomistic models. Effective continuum approaches [14], which use relevant inputs from atomistic simulations, have been developed for integrating atomistic models to continuum models. Ideally, the behavior of an effective continuum closely resembles that of the atomistic structure, under any set of boundary conditions. Quasi-continuum approaches [15] have been established, in which the deformation of an RVE of individual atoms in an atomistic model is mapped into an FE model such that the nodes of the FEM deform in an identical manner as the corresponding points in the molecular model [2]. In addition, the deformation energies for the atomic and FE models are the same for identical loading conditions. Other approaches that have been explored for bridging the atomistic–continuum models are the atomic-scale finite-element method (AFEM) [16], bridging domain method [17], and bridging scale method [18].

7.3
Application of Computational Modeling Techniques to Energetic Polymer Formulations

This section provides details on the application of the various computational modeling techniques to the study of different aspects of energetic binders and their formulations.

7.3.1
Quantum Mechanical Methods

Quantum mechanical based computational approaches have been used to predict the heat of formation of the energetic polymer Poly(3,3-bis(azidomethyl) oxetane) (Poly(BAMO)) and its copolymer with tetrahydrofuran (THF) [19(a)]. Compared with typical C–H–N–O explosives, BAMO is a small energetic molecule. It is a four membered heterocyclic molecule with very high stability. Its high ground state energy and the four membered ring structures are the attractive feature of this system for theoretical studies, particularly as a prototype for modeling the complex physics and chemistry associated with initiation and growth of reactions to detonation in solid energetic materials. As we can easily model the geometry, stereochemistry, and the ground-state energies of the ring systems, these systems can be easily modified with the help of theory and then the synthesis of only the potential candidates will be carried out in the laboratory. Such studies can be used to identify and characterize phenomena that contribute to explosive sensitivity, thereby creating a theoretical means with which to screen energetic materials for insensitive munitions (IM) applications. In this way, theoretical simulations serve to reduce the risk associated with time-consuming and repetitive testing of these hazardous materials.

A natural starting point for theoretical studies of the initiation of the molecules is to investigate chemical decomposition at the molecular scale. For this purpose, *ab initio* quantum chemical calculations provide the most accurate and detailed description of the reaction chemistry in the absence of empirical data. Energetic binders made from polyoxetanes have been investigated for use in gun propellants and in other applications where an energetic binder would be useful. In order to predict the performance of an energetic material, it is important to have accurate, experimentally determined values for the enthalpy of formation (ΔH_f) of the material.

The main aim of this work [19(a)] was to calculate the heat of formation of the BAMO and its copolymer with THF so as to predict the energetic binder network that delivers a better performance to energetic material composites. Different types of polymers (block and alternating) have been designed and the ΔH_f of these polymeric chains were calculated using *ab initio* methods. Ab initio method predicts the copolymerization route to be adopted to obtain an energetic binder with high heat of formation. Furthermore, ab initio methods could be used to compute the vibrational spectra of the designed polymers which holds a direct correlation with the thermal stability of the polymer.

All the molecules have been optimized using restricted Hartree–Fock formalism by using a doubly diffused and doubly polarized 6–31 G basis set available in the Gaussian codes to obtain the maximum sensitivity to the optimized geometry in the prediction, because of the high sensitivity of the frequency values to the geometry. The optimized geometries were used to compute the SCF (self-consistent field) MO (molecular orbital) energies and ΔH_f. The static vibrational frequencies of molecules have been studied with the semi-empirical AM1 level of theory available in the Gaussian codes. The studies were restricted to the AM1 level to overcome the high computational cost of *ab initio* calculation for frequency predictions.

The optimized geometries of the polymers using the HF/6–31++G** basis set is given in Figure 7.5.

As the efficiency of the energetic materials as gun propellants mainly depend on their heats of formation (H_f), H_f values were calculated with a high degree of accuracy using doubly diffused and doubly polarized basis sets using Hartree–Fock formalism. The results are given in Table 7.1.

The calculated value of heat of formation of BAMO is in good agreement with the reported literature experimental values (BAMO: 0.616 kcal/g) [19(b)]. These data prove the accuracy of the theoretical calculation using the *ab initio* methods.

7.3 Application of Computational Modeling Techniques to Energetic Polymer Formulations

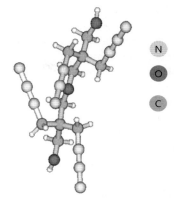

Optimized structure of Poly(BAMO)

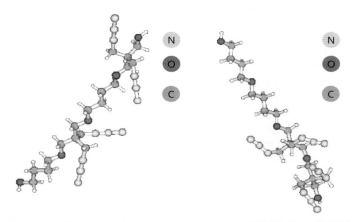

Optimized structure of BAMO-THF alternating polymer

Optimized structure of BAMO-THF block polymer

Figure 7.5 Optimized structures of polymers (two repeating units) [19a].

Table 7.1 Heat of formation of energetic polymer Poly(BAMO) and its copolymers with THF [19a].

Molecule	ΔH_f (kcal/g)[a]
BAMO	0.635
Poly(BAMO)	0.960
BAMO-THF (BD) alternating copolymer	0.810
BAMO-THF (BD) block copolymer	0.083

[a]1 cal=4.184 J.

Vibrational frequencies of polymers are calculated using the semi-empirical AM1 method. As the normal modes tend to be a combination of several types of bond motions (e.g., stretching or bending), identification of mode type is limited to a description of the most significant displacements. Of particular interest are the stretching modes of terminal C–H, N–N, and C–O bonds. The frequencies of these vibrations are the clear indications of thermal stability and/or decompositions of these systems. The low frequencies of the modes indicate that they require low activation energies for bond scissions, and play an active role in the thermal decomposition of the energetic molecules. The results are presented in Table 7.2.

The heats of formation (ΔH_f) and IR absorption frequencies of polymers calculated using *ab initio* and semi-empirical methods, respectively, help to provide information about the thermal decomposition and performance of high-energy material based on Poly(BAMO) derivatives.

Energetic oxetane polymers are synthesized by the cationic ring opening polymerization of the respective oxetane monomers. *Ab initio* quantum mechanical approaches have been used to study the reactivity of oxetane monomers towards ring opening polymerization and also for the analysis of the mechanistic aspects of the polymerization reaction [20]. Furthermore, *ab initio* techniques are also used for the theoretical validation of photoelectron spectra profiles [21] of halogenated methyl oxetane derivatives, which are the precursors for energetic monomers.

The propensity of energetic monomers to polymerize is dependent on their basicities and ring strain. It is desirable to predict the reactivity for ring opening of the energetic monomers prior to their tedious synthesis. Differences in the reactivity among the monomers are primarily due to: (i) the basicity of the O atom and (ii) differences between the HOMO (highest occupied molecular orbital) energy of the oxetanes and the LUMO (lowest unoccupied molecular orbital) energy of the activated oxetane polymeric chains [20(c)].

Ab initio calculations were employed to generate electrostatic molecular potential contour maps to predict differences in basicities among the energetic monomers and to predict their reactivities [20]. Mechanistically, in the propagation step of the ring opening polymerization, the oxetane molecules react with the protonated oxetane species in the initiation step, with concomitant ring opening of the protonated oxetane. Kaufman and coworkers used *ab initio* quantum chemical

Table 7.2 IR frequencies of polymers calculated by QM methods [19a].

Polymer	Stretching modes (wave number in cm^{-1})		
	C–H	N–N	C–O
Poly(BAMO)	1065.21	1524.18	2024.34
BAMO-THF (BD) block polymer	1178.54	1743.09	2315.90
BAMO-THF (BD) alternating polymer	1270.13	1824.15	2528.43

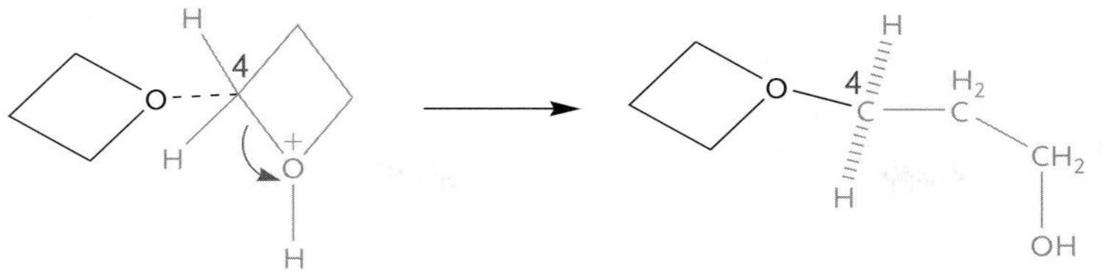

Scheme 7.1 Ring opening mechanism of oxetane predicted by *ab initio* method [20].

model potential/variable retention of diatomic differential overlap (MODPOT/VRDDO) MRD-CI (multiple reference double excitation–configuration interaction) formalism to theoretically investigate the reaction mechanism of the ring opening polymerization of energetic oxetanes [20].

The calculated potential energy surfaces indicated that [20] the reaction pathway resembles an S_N2 reaction, with the oxygen of the oxetane attacking linearly along the C4–O direction of the protonated oxetane and inversion of the hydrogens around the C4 atom (Scheme 7.1).

Ab initio methods have been employed to investigate the decomposition pathways of mixtures of energetic glycidyl azide polymer (GAP) with various nitramines (HMX, RDX, and CL-20) using the B3LYP/3-21G level of theory [22]. The breaking of the N–NO$_2$ bond is considered to be the most energetically favorable initial decomposition step for nitramines. The computations predict that the presence of GAP decreases the activation energy of NO$_2$ elimination by 8 kJ/mol for CL-20, whereas the NO$_2$ elimination from HMX is only favored by 1 kJ/mol, and NO$_2$ elimination from RDX is inhibited in the presence of GAP diol by 2 kJ/mol [22]. This prediction indicates that the decomposition mechanism changes upon addition of GAP, which agrees well with the isolated decomposition products experimentally.

7.3.2
Molecular Dynamics (MD) Simulations

The copolymer of BAMO with 3-azidomethyl 3-methyl oxetane (AMMO) is an energetic thermoplastic elastomer (ETPE), which can be processed as a thermoplastic using melt extrusion techniques. Interfacial slip between the polymer melt and the metal wall of the extruder barrel (made of iron) leads to flow instabilities resulting in the development of surface distortions of the extrudates. It was observed experimentally that, unlike the behavior of homopolymers such as high density polyethylene (HDPE), the copolymers do not exhibit wall slip over relatively high strains/high shear rates. Molecular dynamics were utilized to understand the molecular basics of this behavior of the ETPE [23].

The simulations pointed out that the energy necessary for separating the polymer molecules from the iron surface (adhesive energy value) for the copolymer is greater than twice that of HDPE and polydimethylsiloxane (PDMS), thereby reducing the possibility of wall slip and providing stable flow conditions. The high adhesive energy was attributed to the strong electrostatic interactions between the negatively charged azido groups of BAMO/AMMO and the positively charged iron and hydrogen atoms of the metal wall. Therefore, molecular modeling techniques are able to reveal the molecular scale interactions between the polymer melt and the extrudate barrel and support the experimentally observed rheological behavior [23].

Molecular mechanics and dynamics were utilized to study the interface of a β-HMX/Poly(3-nitratomethyl-methyl oxetane) ((Poly(NIMMO)) formulation cured with MDI (4,4′-diphenylmethane diisocyanate) to understand the filler matrix interactions at a molecular level [24]. The interactions were studied with respect to three perfect crystalline faces of β-HMX, namely (011), (010), and (110). Polarities of the crystalline faces were assessed by their interaction with water. Higher affinity of a given face towards water was attributed to increasing oxygen atom population on the surface of the face (surface polarity).

As illustrated in Figure 7.6, for both Poly(NIMMO) and MDI, the weakest interaction occurred with the most polar face, (110) of β-HMX. This behavior was attributed to the electrostatic characteristics of MDI and Poly(NIMMO), which were dominated by the electronegative oxygen atoms of the model molecules. Such information will help to design high performance formulations of plastic bonded explosives (PBX) using energetic polymers [24].

MD simulation methods have been used to investigate the mechanical properties of PBX formulations of ε-CL-20, HMX, and TATB (triaminotrinitrobenzene) with four types of fluoropolymers: poly(vinylidene difluoride) (PVDF), polychlorotrifluoroethylene (PCTFE), F2311, and F2314 [25]. F2311 and F2314 are copolymers polymerized from vinylidene difluoride and chlorotrifluoroethylene with molar ratios of 1 : 1 and 1 : 4, respectively. The mechanical properties were

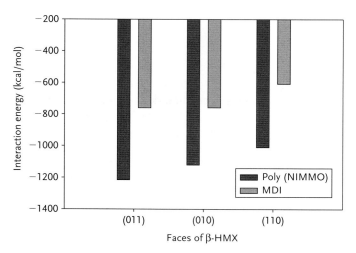

Figure 7.6 Plot showing the interaction energies of Poly(NIMMO) and MDI with three different crystal faces of β-HMX [24].

7.3 Application of Computational Modeling Techniques to Energetic Polymer Formulations

assessed with respect to different crystalline surfaces of the explosives [(001), (100), (010)] using the advanced COMPASS (condensed-phase optimized molecular potentials for atomistic simulation studies) force field.

It was observed that, in general, the mechanical properties of the explosives could be effectively improved by blending them with polymer binders in a small amount. Different mechanical properties were obtained when the polymers were simulated to contact different crystalline faces of the explosives. For TATB, the whole effect of improving the mechanical properties was found to be (010) $\approx$ (100) > (001), and for ε-CL-20, (100) $\approx$ (001) > (010) [25(b and c)].

The binding energy of polymers with the different crystalline faces of ε-CL-20 was evaluated using MD simulations [25(a)]. Among the polymers, F2314 had the highest binding energy with the different crystal faces of ε-CL-20, and the ordering was observed to be (001) > (100) > (010). Physically, when the fluoropolymers are blended with ε-CL-20, they have a tendency to concentrate on the (001) face, which has the largest binding energy [25(a)]. Thus MD simulations are a valuable tool for experimentalists to choose the best combination of binders and explosives for a given practical application.

MD simulations have been used to predict the thermodynamic transport and viscoelastic properties of polymer binders used for PBX composites using a previously developed link between MD simulations and well established theories [26].

Smith and coworkers [26(a)], using a quantum chemistry based potential force field, predicted the variation of volume and bulk modulus of Estane (binder for HMX based explosive, PBX 9501) with respect to pressure at isothermal conditions. The predictions were found to be in excellent agreement with the empirical equation of states. Furthermore, variations of density, apparent viscosity, self-diffusion coefficients, and viscoelastic properties of polymer binders were accurately predicted by MD simulations. The simulations were found to be in excellent agreement with the experimental results (Figure 7.7). Similar MD simulations can be extended to the new series of energetic polymer binders to understand their properties, which will help in the formulation of PBX compositions with enhanced performance.

A novel approach for investigating reactions in the condensed phase is to take advantage of the computational efficiency of empirical force fields. Traditional empirical force fields are unable to model bond-breaking events such as thermal decomposition. Additional force fields dependent on the bond order are included in the overall force field. These bond-order terms semi-empirically describe the bond-breaking mechanisms needed to simulate the reaction conditions. Goddard and coworkers [27] developed a method called ReaxFF, which incorporates non-bonded interactions and coulombic forces along with the central-force formalism to describe smooth dissociations between atoms. Local perturbations (torsion, bond angle) are added for accurate descriptions of complex molecules.

In PBX formulations, the main purpose of binding polymers to high explosives is to reduce the sensitivity to shock, friction, and impact. ReaxFF has been used to study the thermal decomposition of RDX in the presence and absence of Estane chains [27]. The simulation system consists of 64 RDX molecules and two Estane

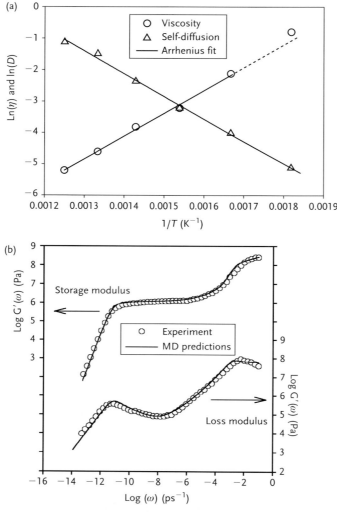

Figure 7.7 MD predictions of variation of: (a) self-diffusion coefficients, viscosity of Estane binder [26(a)]; and (b) viscoelastic properties of polybutadiene [26(b)]. Re-printed by permission of Taylor & Francis Group [Figure 26(a)] and American Chemical Society, Copyright (2002) [Figure 26(b)].

chains. The total chemical fragments produced during the cook-off simulations for pure RDX and RDX–Estane were measured at picosecond intervals. The total number of fragments for pure RDX is always larger than that for the RDX–Estane, indicating that the polymer binder Estane reduces the thermal decomposition process of pure RDX. The pressure of pure RDX (9.9 GPa) is higher than that for the RDX–Estane because more fragments are generated in the cook-off processes for pure RDX than that for the PBX. These results show that when the Estane chains are bonded to the RDX crystal, the decomposition of RDX crystal decreases, indicating that the sensitivity of RDX to thermal conditions is reduced. Thus the ReaxFF based MD simulation approach could quickly test the response of PBX to thermal conditions and predict the sensitivity [27].

7.3.3
Mesoscale Simulations

All energetic materials are heterogeneous in nature and consist of a mixture of polycrystalline and binder materials. The energetic composites exhibit distinctly different behaviors as compared with the pure materials. The properties could be ascribed to the "mesoscale effects" – the scale associated with the granular nature of the constituents of the energetic composite [28]. For example, the threshold to reaction of a PBX is greatly influenced by its morphology, defect content, and particle size distribution, which are all characterized by mesoscale dimensions. Hence simulations at the mesoscale are critical for prediction of properties of PBX.

Mesoscale simulations have been employed to understand the mechanism of irreversible growth behavior (Ratchet growth) exhibited by PBX formulations consisting of explosives with a layered crystal structure. PBX-9502 is a TATB-based PBX showing irreversible crystal growth [29]. It is the explanation for a significant difference in dimensional measurements before and after thermal hysteresis. This phenomenon does not occur in single crystals of TATB, but is a complex product of interactions of TATB particles among themselves and with the polymer binder [29(c)]. The mesoscale simulations conclusively demonstrates that irreversible growth happens only if the crystalline explosive is of anisotropic nature and the phenomenon is mediated by explosive particles much smaller than the average crystallite size [29(a)]. Furthermore, the simulations demonstrate that the irreversible growth phenomenon depends on the nature of the binder, and it can be substantially reduced by using a fluoropolymer (Cytop) with a glass transition temperature higher than the thermal cycle maximum as the binder for PBX [29(b)].

Mesoscale simulations based on dissipative particle dynamics (DPD) have been used to investigate the influence of chain length and temperature on the phase separation behavior between polymer binder and plasticizer in nitrate ester plasticized polyether propellant [30]. The final morphology of the propellant depends on the competition between thermodynamic and kinetic factors. The mesoscopic simulations suggest that temperature is a key factor determining the phase separation between plasticizer and the binder. At lower temperatures, the phase separation takes place in later stages of the simulation and the degree of phase separation was extensive. The longer the polymer chain, the earlier the phase separation takes place during the initial stages of simulation. The extent of phase separation decreases with increasing length of the polymer chain.

The initiation and reactivity of the energetic composite is greatly influenced by the mesoscopic microstructure arising from the heterogeneous nature of the composite material. Although the energy release from the energetic composite is taking place at the molecular level, its transformation to other forms of energy (thermal and mechanical) takes place through the mesoscopic microstructure, known generally as "mesoscopic effects" [28]. Mesoscopic modeling was adopted to reveal the mesoscale response of a PBX and its constituents (HMX crystals, nitroplasticized Estane binder, and a mixture of fine HMX crystallites and binder (called dirty binder)) subjected to a ramped isentropic loading experiment [31]. An isentropic compression experiment (ICE) suppresses the energy localization effects and provides insights into the material behavior without the complexity of any chemical reactions. The details of the simulation indicate that much of the plastic deformation first occurs at the large HMX crystal contact points. Thereafter, the large HMX crystals are confined and the subsequent deformation of the composite is controlled by the dirty binder materials in the interstitial sites of the composite. Initiation of the energetic composite is triggered by the energy localization points (hot spots) caused due to the localized plastic deformation of the large HMX crystals.

7.3.4
Macroscale Simulations

Macroscale simulations are mainly focused on the prediction of mechanical performance of the PBX compositions. The main challenge in macroscale simulations is the identification of a theoretical model that captures the complex non-linear mechanical response of PBX. The high solids loading present in PBXs makes the mechanical response of PBXs difficult to predict, as the composites show some characteristics of polymer binders and some of the explosive solids, each of these phases exhibiting a complex constitutive response varying with temperature and load. The mechanical characterization of the overall constitutive response of the PBX is challenging, because it requires the coupling of the deformation of constituent phases in entirely different regimes, namely, elastic and viscous [32]. Furthermore, the problem becomes more challenging if the dewetting (interfacial debonding) of the explosive particulates from the polymer binder is taken into account, leading to the formation of interfacial cracks or holes. This microstructural feature is the primary reason for the material to exhibit a strong non-linear response.

Fracture of PBX is primarily governed by the interfacial debonding between the polymer binder and energetic particles [33]. A continuum damage model, where the fracture is modeled as the limit of a process of damage localization in continuum, is inadequate to represent the interfacial debonding behavior of PBX. Discrete damage models based on cohesive laws, which are capable of incorporating the discontinuities in materials, have been used to study the interfacial debonding process in composite materials [34]. Cohesive laws provide a relationship between the stress across the failure process zone and the opening displacement. For PBX composites, the interfacial debonding between the polymer binder and the energetic particulates are rate dependent due to the viscoelastic nature of the polymer binder. Therefore, in order to develop accurate models for the failure of PBX, viscoelastic or rate dependent cohesive zone models [35] should be implemented.

Other than the interfacial debonding, energetic filler particles also play a dominant role in determining the failure behavior of PBX. The energetic particles undergo crystal fracture, especially at higher compressions due to impacts, leading to the formation of vacuoles [33]. Moreover, they elicit different deformation responses under various loading conditions. For example, HMX behaves as an elastic solid under tension, but is elastic–plastic under compression. A continuum-damage approach based on an isotropic viscoplastic model [36] is required for modeling of the deformation of energetic crystals in the PBX. Thus, the continuum and discrete approaches should be integrated to develop an accurate and robust model [35]. Once the model is developed, it is implemented in an FEM code to carry out the macroscale simulation and predict the mechanical behaviors of PBX.

7.4
Future Outlook

The ultimate properties of energetic materials are determined by complex physio-chemical processes taking place in multiple time and length scales. It is difficult to predict the properties with accuracy by using traditional modeling and simulation methods on a single length and time scale. Multiscale simulation strategies should be adopted to bridge the models and simulation techniques across a broad range of length and time scales in order to link the macroscopic or mesoscopic behaviors of materials to the molecular or atomic scale description. The challenge in multiscale simulation is to move from one scale to another with out any flaws, so that calculated parameters can be efficiently transferred from one scale to another.

Application of multiscale simulation methodologies to energetic materials research is in its infancy. As illustrated, the status of research in modeling and simulation of energetic composites are largely limited to each time and length scales. It is worth noting that some efforts have recently been put into the multiscale strategies for energetic composites. Matous and *et al.* [37] explored multiscale modeling using the mathematical theory of homogenization (MTH) to understand the damage mechanics of energetic composites. Their objective was to relate the micro/meso-scale damage (particle debonding) to the macroscopic constitutive response of the material. However, suitable simulation methodologies have yet to be developed to correlate the atomic/molecular structure of the material to the mesoscopic properties, such as morphology, which determine the ultimate macroscopic properties of the material. Multiscale methods are deemed to gather more importance in the emerging area of nanoenergetic materials, where length and time scales range from Angstroms and femtoseconds for the vibrations of atomic bonds to millimeters and seconds for the crack propagation in the composites.

References

1 Pantelides, S.T. (1994) Frontiers in computational material science, *Comp. Mater. Sci.*, **2**, 149–155.
2 Gates, T.S., Odegard, G.M., Frankland, S.J.V., and Clancy, T.C. (2005) Computational materials: multi-scale modeling and simulation of nanostructured materials. *Compos. Sci. Technol.*, **65**, 2416–2434.
3 (a) Chakraborty, D., Muller, R.P., Dasgupta, S., and Goddard, W.A. III, (2000) The mechanism for unimolecular decomposition of RDX, an *ab initio* study. *J. Phys. Chem.*, **101**, 2261–2272; (b) Okovytyy, S., Kholod, Y., Qasim, M., Fredrickson, H., and Leszczynski, J. (2005) The mechanism of unimolecular decomposition of CL-20. A computational DFT study. *J. Phys. Chem. A*, **109**, 2964–2970; (c) Rice, B.M., Byrd, E.F.C., and Mattson, W.D. (2007) *Struct. Bond*, **125**, 153–194.
4 (a) Rappaport, D.C. (2004) *The Art of Molecular Dynamics Simulation*, 2nd edn, Cambridge University Press, Cambridge; (b) Leach, A.R. (2001) *Molecular Modeling Principles and Applications*, Prentice Hall, Englewood Cliffs, NJ.
5 Car, R. and Parrinello, M. (1985) Unified approach for molecular dynamics and density functional theory. *Phys. Rev. Lett.*, **55**, 2471–2474.
6 (a) Fermeglia, M. and Pricl, S. (2007) Multiscale modeling for polymer systems of industrial interest. *Prog. Org. Coat.*, **58**, 187–199; (b) Kremer, K. and Muller-Plathe, F. (2001) Multiscale problems in polymer science: simulation approaches, *MRS Bull.*, **26**, 169.
7 Lopez, C.F., Moore, P.B., Shelley, J.C. and Klein, M.L. (2002) Computer simulation studies of bio-membranes using a coarse grain model. *Comput. Phys. Commun.*, **147**, 1–6.
8 (a) Fraaije, J.G.E.M., van Flimmeren, B.A.C., Maurits, N.M., Postma, M., Evers, O.A., Hoffmann., C. Altevogt, P., Zvelindovsky, A.V., and Goldbeck-Wood, G. (1997) The dynamic-field density functional method and its application to the mesoscopic dynamics of quenched block copolymer melts. *J. Chem. Phys.*, **106**, 4260–4269; (b) Hoogerbrugge, P.J. and Koelman, J.M.V.A. (1992) Simulating microscopic hydrodynamic phenomena with dissipative particle dynamics. *Europhys. Lett.*, **19**, 155–160.
9 (a) Baschnagel, J., Binder, K., Doruker, P., Gusev, A.A., Hahn, O., Kremer, K., Mattice, W.L., Paul, W., Santos, S., Suter, U.W., and Tries, V. (2000) Bridging the gap between atomistic and coarse-grained models of polymers: status and perspectives. *Adv. Polym. Sci.*, **152**, 41–156; (b) Muller-Plathe, F. (2002) Coarse-graining in polymer simulation: from the atomistic to the mesoscopic and back. *Chem. Phys. Chem.*, **3**, 754–769.
10 Valavala, P.K. and Odegard, G.M. (2005) Modeling techniques for determination of mechanical properties of polymer nanocomposites. *Rev. Adv. Mater. Sci.*, **9**, 34–44.
11 Zeinkiewicz, O.C. and Taylor, R.L. (2000) *The Basis*, Vol. **1**, The Finite Element Method, Butterworth-Heinemann, Oxford.
12 Banerjee, P.K. (1994) *The Boundary Element Methods in Engineering*, Mc-Graw Hill, London.
13 (a) Baeurle, S.A. Multiscale modeling of polymer materials using field theoretic methodologies: a survey about recent developments. *J. Math. Chem.*, doi: 10.1007/s 10910-008-9467-3; (b) Baeurle, S.A., Fredrickson, G.H., and Gusev, A.A. (2004) Prediction of elastic properties of poly (styrene-butadiene-styrene) using a mixed finite element approach. *Macromolecules*, **37**, 5784–5791.
14 (a) Lee, U. (1994) Equivalent continuum models for large plate-like lattice structures. *Int. J. Solids Struct.*, **31** (4), 457–467; (b) Odegard, G.M., Gates, T.S., Nicholson, L.M., and Wise, K.E. (2002) Equivalent continuum modeling of nano-structured materials. *Compos. Sci. Tech.*, **62**, 1869–1880.
15 (a) Tadmor, E.B., Ortiz, M., and Phillips, R. (1996) Quasicontinuum analysis of

defects in solids. *Philos. Mag. A*, **73**, 1529–1593; (b) Miller, R.E. and Tadmor, E.B. (2002) The quasicontinuum method: overview, applications and current directions. *J. Comput. Aided Mater. Des.*, **9**, 203–239.

16 Liu, B., Huang, Y., Jiang, H., Qu, S., and Hwang, K.C. (2004) The atomic-scale finite element method. *Comput. Meth. Appl. Mech. Eng.*, **193**, 1849–1864.

17 Xiao, S.P. and Belytschko, T. (2004) A bridging domain method for coupling continuum with molecular dynamics. *Comput. Meth. Appl. Mech. Eng.*, **193**, 1645–1669.

18 Wagner, G.J. and Liu, W.K. (2003) Coupling of atomistic and continuum simulations using a bridging scale approximation. *J. Comput. Phys.*, **190**, 249–274.

19 (a) Maheshkumar, M.V., Joseph, M.J., Sreekumar, K., and Ang, H.-G. (2006) Synthesis and characterization of poly (BAMO) suitable for binder application. *Chin. J. Energ. Mater.*, **14** (6), 411–415; (b) Schmidt, R.D. and Manser, G.E. (2001) Heat of formation of energetic oxetane monomers and polymers. Proceedings of 32nd International Annual Conference of ICT, Karlsruhe.

20 (a) Kaufman, J.J., Hariharan, P.C., and Keegstra., P.B. (1987) *Ab initio* MRD-CI calculations for the propagation step of cationic polymerization of oxetanes based on localized orbaitals. *Int. J. Quantum Chem., Quantum. Chem. Symp.*, **21**, 623–643; (b) Kaufman, J.J., Hariharan, P.C., Roszak, S., and Keegstra, P.B. (1987) *Ab Initio* MRD-CI calculations on protonated cyclic ethers: (1) protonation pathways involve multi potential surfaces, (2) differences from SCF in dominant configurations upon opening epoxides. *Int. J. Quantum Chem., Quantum. Biol. Symp.*, **14**, 37–46; (c) Cheun, Y.-G. and Cho, S.G. (1995) A study on electrostatic potentials and chemical reactivities of energetic oxetanes. *J. Kor. Chem. Soc.*, **39**, 329–337.

21 Roszak, S., Kaufman, J.J., Koski, W.S., Barreto, R.D., Fehlner, T.P., and Balasubramanian, K. (1992) Experimental and theoretical studies of photoelectron spectra of oxetane and some of its halogenated methyl derivatives. *J. Phys. Chem.*, **96**, 7226–7230.

22 Bohn, M.A., Hammerl, A., Harris, K., and Klapotke, T.M. (2005) Interactions between the nitramines RDX, HMX and CL-20 with the energetic binder GAP. Proceedings of the VIII Seminar on "New Trends in Research of Energetic Materials" University of Pardubice, Pardubice, Czech Republic, April 19–21.

23 Olgun, U. and Kalyon, D.M. (2005) Use of molecular dynamics to investigate polymer-melt wall interactions. *Polymer*, **46** (22), 9423–9433.

24 Cumming, A.S., Leiper, G.A., and Robson, E. (1993) Molecular modeling as a tool to aid the design of polymer bonded explosives. Proceedings of the 24th International Annual Conference of ICT, Karlsruhe.

25 (a) Xu, X.J., Xiao, H.M., Xiao, J.J., Zhu, W., Huang, H., and Li, J.S. Molecular dynamics simulation of pure ε-CL-20 and ε-CL-20 based PBXs. *J. Phys. Chem. B*, **110**, 7203–7207 (2006); (b) Xiao, J.J., Ma, X., Zhu, W., Huang, Y., and Xiao, H.M. (2007) Molecular dynamics simulations of polymer bonded explosives (PBXs): modeling, mechanical properties and their dependence on temperatures and concentration of binders. *Propellants Explos. Pyrotech.*, **32** (5), 355–359; (c) Xiao, J.J., Huang, Y. Hu, Y., and Xiao, H.M. (2005) Molecular dynamics simulations of mechanical properties of TATB/fluorine-polymer PBXs along different surfaces. *Sci. China B Chem.*, **48** (6), 504–510; (d) Xiao, H.M., Li, J.S., and Dong, H.S. (2001) A quantum chemical study of PBX: intermolecular interactions of TATB with CH_2F_2 with linear fluoro containing polymers. *J. Phys. Org. Chem.*, **14**, 641–649.

26 (a) Davande, H., Bedrov, D., and Smith, G.D. (2008) Thermodynamic, transport, viscoelastic properties of PBX-9501 binder. A MD simulation study. *J. Ener. Mater.*, **26** (2), 115–138; (b) Byutner, O. and Smith, G.D. (2002) Viscoelastic

properties of polybutadiene in the glassy regime from molecular dynamic simulations. *Macromolecules*, **35**, 3769–3771.

27 Zhang, L., Zybin, S.V., van Duin, A.C.T., Dasgupta, S., and Goddard, W.A. III (2005) Thermal decomposition of energetic materials by REAXFF reactive molecular dynamics. 14th American Physical Society Topical Conference on Shock Compression of Condensed Matter, 589–592.

28 Baer, M.R. (2002) Modeling heterogeneous energetic materials at the mesoscale. *Thermochim. Acta*, **384**, 351–367.

29 (a) Gee, R.H., Maiti, A., and Fried., L.E. (2007) Mesoscale modeling of irreversible volume growth in powders of anisotropic crystals. *Appl. Phys. Lett.*, **90**, 254105; (b) Maiti, A., Gee, R.H., Hoffman, D.M., and Fried, L.E. (2008) Irreversible volume growth in polymer-bonded powder systems: effects of crystalline anisotropy, particle size distribution and binder strength. *J. Appl. Phys.*, **103**, 053504; (c) Skidmore, C.B., Butler, T.A., and Sandoval, C.B. (2003) The elusive coefficients of thermal expansion of PBX 9502. LANL Technical Report LA-14003, Los Alamos National Laboratory.

30 Li, S., Liu, Y., Tuo, X., and Wang., X. (2008) Mesoscale dynamic simulation on phase separation between plasticizer and binder in NEPE propellants. *Polymer*, **49**, 2775–2780.

31 Baer, M.R., Hall, C.A., Gustavsen, R.L., Hooks, D.E., and Sheffield, S.A. (2007) Isentropid loading experiments of a PBX and constituents. *J. Appl. Phys.*, **101**, 034906.

32 Xu, F., Aravas, N., and Sofronis, P. (2008) Constitutive modeling of solid propellant materials with evolving microstructural damage. *J. Mech. Phys. Solids.*, **56** (5), 2050–2073.

33 Rae, P.J., Goldrein, H.T., Palmer, S.J.P., Field, J.E., and Lewis, A.L. (2002) Quasi-static studies of the deformation and failure of the β-HMX based PBX. *Proc. R.. Soc. London, A Math. Phys. Sci.*, **458** (2019), 743–762.

34 Tan, H., Liu, C., Huang, Y., and Guebelle, P.H. (2005) The cohesive law for the particle/matrix interfaces in high explosives. *J. Mech. Phys. Solids*, **53** (8), 1892–1917.

35 Wu, Y.-Q. and Huang, F.-L. (2009) A micromechanical model for predicting combined damage of particles and interface debonding in PBX explosives. *Mech. Mater.*, **41**, 27–47.

36 Menikoff, R. and Thomas, D.S. (2002) Constituent properties of HMX needed for mesoscale simulations. *Combust. Theor. Model*, **6** (1), 103–125.

37 Matous, K., Inglis, H.M., Gu, X., Rypl, D., Jackson, T.L., and Geubelle, P.H. (2007) Multiscale modeling of solid propellants: from particle packing to failure. *Compos. Sci. Tech.*, **67**, 1694–1708.

Index

Page numbers in *italics* refer to figures and tables.

3M Company 10
ABA ETPE systems 122
Abel, Sir Frederick 2
Acetyl nitrate 88
Acid catalysts
– for polymerization of NIMMO 88
– for polymerizations of GN to PGN 84
Acroyl chloride (AC)
– curing systems based on 60
Activated monomer mechanism
  (AMM) 83–84
Active chain end (ACE) 83–84
Aerojet 4, 5
Al/PTFE composite 149–52
Aliphatic diisocyanate curing agents 36
Alkoxide anion 30
Allophanate network formation 35
Aluminized explosives 65
Aluminum 3
Ammonium dinitramide (ADN)
  propellants 103–5, 107
Ammonium nitrate (AN) 57
Ammonium perchlorate (AP) propellants
  59, 101–3
Anionic polymerization 4
Aromatic curing agent 36
Aromatic isocyanates 27
Arrhenius parameters
– for different azido polymers 45
– for different types of GAP 46
– for HTPB 46
Arrhenius plots
– for azido polymers 45
Atomic Weapons Establishment (AWE) 11
Atomic-scale finite-element method
  (AFEM) 197
Azide terminated glycidyl azide (GAPA) 175

Azido acetate ester based plasticizers 172–74
Azido based oligomeric plasticizers 174–78
Azido ester plasticizers 179
– physical properties of *173*
– thermal and sensitivity properties of *173*
Azido group plasticizers 172–78
Azido groups, decomposition of 132–34
3-Azidomethyl 3-methyl oxetane
  (AMMO) 10, 24, 201
– burning rates of 137
– copolymers of 132
Azido polymers
– as explosive binders 63–66
– based PBX formulations for underwater
  explosives 65–66
– based propellants performance of 61–63
– burn rate of 49
– combustion of 47–49
– curing of 25–36
– GAP 62
– preparation of
– – from oxetanes 22–24
– – Glycidyl Azide Polymer (GAP) 19–22
– thermal decomposition and combustion
  of 49–61
– – and ammonium nitrate composite
  propellants 57–58
– – and nitramine mixtures 50–57
– – propellants with HNF 58–61
– – thermal decomposition characteristics of
– – kinetics of 43–47
– – mechanism of 41–43
– with inert HTPB, comparative properties
  of *25*
Azido terminated GAP 13
– chemical structure of *13*
Aziridines 3

Energetic Polymers: Binders and Plasticizers for Enhancing Performance, First Edition.
How Ghee Ang and Sreekumar Pisharath.
© 2012 WILEY-VCH Verlag GmbH & Co. KGaA, Weinheim. Published 2012 by WILEY-VCH Verlag GmbH & Co. KGaA

Ballistite 2
BAMO-AMMO
– formulations 138
– thermal decomposition of 132
BAMO-AMMO copolymer, combustion of
– effect of AMMO segments on 136
BAMO-AMMO ETPE
– decomposition of 135
– propellants of 140
BAMO-GAP-BAMO ETPE, preparation
  of 125
BAMO-NIMMO ETPE
– synthesis of, using block-linking
  approach 123
– thermal decomposition of 132
BAMO-NIMMO-BAMO ETPE
– synthesis of 129
Bartley, Charles 3
BDNPA/F plasticizer 187
– properties of 187
– structure of 186
Binders, for high explosives 5
bis(2,2-dinitropropyl) acetal (BDNPA) 36
bis(2,2-dinitropropyl) diformal
  (BDNPDF) 188
bis(2,2-dinitropropyl) formal (BDNPF) 36,
  188
p-bis(α,α-dimethylchloromethyl) benzene
  (p-DCC) 127
1,3-bis(azidoacetoxy)-2,2-bis(azidomethyl)
  propane (P2) 172
– chemical structures of 172
1,3-bis(azidoacetoxy)-2-azidoacetoxymethyl-2-
  ethylpropane (P1) 172
– chemical structures of 172
bis(azido acetoxy) bis(azido methyl) propane
  (BABAMP) 179
3,3-bis(azidomethyl) oxetane (BAMO) 10, 23,
  122, 201
– copolymers of 132
3,3-bis(chloromethyl) oxetane (BCMO) 10, 23
Boron trifluoride (BF_3)-etherate 23, 24
Bottger, Rudolf Christian 1
Boundary element (BE) 197
Braconnot, Henri 1
British Defense Research Agency (DRA) 9
Brown, E.A. 2
Bubble energy 65
Burn rate
– for polymers 49
– of GAP/AN formulation 58
– of GAP/HMX propellant 56
– of GAP/RDX mixture 56
Burn-rate behavior of GAP
– with dipolarophile and isocyanate 38

Butadiene 3
Butanetriol trinitrate (BTTN) 8, 56, 180
Butarez CTL 4

Carbon tetrachloride 24
Carboxyl terminated polybutadiene
  (CTPB) 4, 13
– chemical structure of 4
ClMMO 24
Coarse-grain (CG) modeling 195
Combustion wave structure, of the GAP/
  nitramine propellant mixture 54
Common polyisocyanates 27
COMPASS (condensed-phase optimized
  molecular potentials for atomistic
  simulation studies) 203
Composite modified double-base (CMDB)
  propellants 100–5
Computational techniques application, to
  energetic polymers and formulations
  193–207
– macroscale simulations 206–7
– mesoscale simulations 205–6
– molecular dynamics (MD)
  simulations 201–5
– overview of 193–207
– quantum mechanical methods 197–201
Continuum method 196
Cordite 2
CPX 413 112
CPX series formulations 112
Curing for GAP and HTPB 29
Curing reaction, of GAP 20
Curing systems, for PBAN 3
Curing, of azido polymers 25–36
– by polyisocyanates 25–36
–– cured polyurethane elastomer, structure
  of 28–29
–– Gel-time characteristics 31–32
–– post-Cure properties 33–36
–– preparation of 26–28
– kinetics of 29–30
– polyisocyanate compounds used for 27
Cyclodextrins (CDs) 89

Dark zone 55
DEGBAA (diethylene glycol bis(azidoacetate)
  ester) 172
– chemical structures of 188
DEGDN (diethylene glycol dinitrate) 189
Dendrimers 177
Dendritic azido ester plasticizer
– chemical structure of 177
Desensitization 6
Desmodur N-100 29

Dewar, James 2
1,1-diamino-2,2-dinitroethylene (FOX-7) 112–13
1,3-di(azidoacetoxy)-2,2-di(azidomethyl) propane (PEAA) 172
2,2-dinitro-1,3-propanediol diformate (ADDF) plasticizers 184
– structure of 184
Di- or tri-functional epoxides 3
4,4′-diaminodiphenylmethane (DDM) 69
Dibutyl phthalate (DBP) 177, 179
Dibutyltin dilaurate (DBTDL) 28, 30
Differential scanning calorimetry (DSC) 43
Differential thermal analysis (DTA) 43
Difluoroamine based energetic binders 160
Difluoroamine polymers
– properties of 160
Di-functional hydroxyethyl tetrazole 67
Dimethylformamide (DMF) 23
Dinitrogen pentoxide (N_2O_5) 9
Dioctyl phthalate (DOP) 179
4,4′-diphenylmethane diisocyanate (MDI) 7, 202
Dissipative particle dynamics (DPD) 196, 205
DSC (differential scanning calorimetry) 178

Effective continuum approaches 197
EGBAA (ethylene glycol bis(azidoacetate) ester) 172
– chemical structures of 188
Electron spin resonance (ESR) 43
Empirical force fields 203
Emulsion radical copolymerization 3
Energetic bisdifluoroamine polymer
– preparation of 159
Energetic filler particles 207
Energetic mono- and bis-difluoroamino-monomers 159
Energetic oxetane polymers 200
Energetic plasticizers 58, 171–89
– azido group plasticizers 172–78
– – azido acetate ester based plasticizers 172–74
– – azido based oligomeric plasticizers 174–78
– azido plasticizers 178–79
– miscellaneous plasticizers based on nitro-Groups 186–89
– – nitratoethylnitramine (NENA) 188–89
– – polynitro-aliphatic plasticizers 186–88
– nitrate ester oligomers 184–85
– nitrate ester plasticizers 180–84
Energetic polyformal 161
– preparation of 161

Energetic polyglycidyl nitrate 13
Energetic polymer plasticizers 12–13
Energetic polymers (other than NC) as binders 8–12
– energetic polyoxetanes 10–11
– – chemical structure of 10–11
– energetic thermoplastic elastomers 11–12
– GAP 9–10
– polyglycidyl nitrate 8–9
– polyphosphazenes 11
Energetic polyoxetanes 13
Energetic salts, of nitrogen heterocycles 68
Energetic thermoplastic elastomers (ETPE) 11–12, 121–44, 201
– burn-rate profiles of 136
– ETPE-based polymer nanocomposites 142–44
– ETPE-based propellant formulations, performance of 139–41
– melt-cast explosives based on 141–42
– preparation of 122–31
– – block-linking approach 123–25
– – sequential addition approach 123, 125
– propellant formulations, combustion of 137–39
– thermal decomposition and combustion of
– – effects of ageing on 132
– thermal decomposition and combustion of 132–37
Energetic thermoplastic elastomers (ETPEs)
– chemical structures of 11, 12
Epichlorohydrin (ECH) 19, 20
– polymerization of 9
Epon X-801 3
N–N-bonded epoxy binders 69–70
Epoxy curing agent 69
ESTANE 5
Estane 5703 7, 122
– chemical structure of 7
Ethylene glycol dinitrate (EGDN) 180
ETPE layered propellant formulations 141
ETPE/Al composites, developing 143
European Space Agency (ESA) 60
Explosive binders 5–6
– azido polymers as 63–66
– chemical structure of 6

Fast pyrolysis techniques 43
Field based dynamic mean field theory 196
Finite-element analysis (FEA) 196
Fluorocarbon polymers 163
Fluoropolymer based explosives
– detonation properties of 163

Fluoropolymers 6
– chemical structure of 6
Fluoropolymers as binders 147–64
– copolymers of tetrafluoroethylene 152–59
– – Kel-F800 153–56
– – Viton A 156–59
– energetic polymers containing fluorine 159–60
– miscellaneous energetic fluoropolymers 161–62
– PBX formulations with fluoropolymers 162–64
– poly(tetrafluoroethylene) 148–52
– – energetic composites of 149–52
– – phase transitions of 148–49
FOF5 113
Fourier transform infrared (FTIR) 52
FOX-7 112–13
Free-radical polymerization 4
Free-volume theory 175

GAP 5527 polyol 10
GAP based plasticizer propellants 179
GAP formulations 138
GAP macro-initiator, chemical structure of *131*
GAP polymer, general structure of *21*
GAP triol plasticizer 175
– properties of *175*
GAP/AN formulation 58
GAP/HNF formulations 59
GAP/HNF propellant burning 59
GAP/TEX mixture 51
– thermal decomposition of *53*
GAPA plasticizer *175*
– properties of *175*
GAP-Poly(BAMO) ETPE
– decomposition of 134–35
Gas chromatography (GC) 43
Glass transition temperature 36
Glycidyl azide (GA) 9
Glycidyl azide Polymer (GAP) *13*, 19–22, 201
– linear and branched, comparative properties of *21*
Glycidyl azide prepolymer 9–10, 172
– chemical structure of *9*
Glycidyl nitrate 8, 13
– polymerization of
– – reaction scheme for *83*
– preparation of
– – using a nitrating mixture *82*
– – using dinitrogen pentoxide *82*
– – using from glycerin *82*
Green solvents 68

Guanidine-5,5'-azotetrazolate (GZT) 109, 110
Gun and rocket propellants 50

Hard polymers, in PBX formulations, 6
Hartree–Fock formalism 198
HC-434 4
HCN 42
Henderson, Charles B. 3
Hexanitrohexaazaisowurtzitane (CL-20) 103
High density polyethylene (HDPE) 201
High nitrogen content polymers 19–70
– azido polymers
– – as explosive binders 63–66
– – based propellants performance of 61–63
– – combustion of 47–49
– – curing by dipolarophiles 38–40
– – curing of 25–36
– – physical properties of 25
– – preparation of 19–24
– – thermal decomposition and combustion of 49–61
– – thermal decomposition characteristics of 41–47
– N–N-bonded epoxy binders 69–70
– tetrazole polymers and their salts 66–69
High performance propellants 62
Historical perspective, polymers as binders and plasticizers 1–13
– energetic polymer plasticizers 12–13
– energetic polymers (other than NC) as binders 8–12
– – energetic polyoxetanes 10–11
– – energetic thermoplastic elastomers 11–12
– – GAP 9–10
– – polyglycidyl nitrate 8–9
– – polyphosphazenes 11
– explosive binders 5–6
– hydroxy terminated polybutadiene 5
– nitrocellulose 1–2
– polybutadienes (PBAA, PBAN and CTPB) 3–4
– polysulfides 2–3
– polyurethanes 4–5
– thermoplastic elastomers 6–8
HMX 100–1, 163
β-HMX 202
HOMO (highest occupied molecular orbital) 200
HTPB/HNF propellant 59
Hydrazinium nitroformate (HNF) propellants 58, 103–5
Hydroxy terminated polybutadiene (HTPB) 5, *13*, 19, 36
– chemical structure of *5*

Hydroxyl terminated azido polymers 26
3-hydroxy methyl-3-methyl oxetane (HMMO) 10, 24
Hydroxy-terminated azido prepolymer 9
Hydroxy-terminated poly(2-fluoro-2,2-dinitroethyl) polynitroorthocarbonate prepolymers 161

Insensitive munition (IM) 5, 65, 112, 198
IPDI (isophorone diisocyanate) 27
Isentropic compression experiment (ICE) 206
Isocyanates 21
– for HTPB 5

K-10 plasticizer 185
Kel-F800 147, 153–56, 163
– chemical structure of 153
– dynamic behavior of 153–55
– PBX 164
– thermal decomposition 155–56
Klager, Karl 5
Kraton 121
Kraton G-6500 7, 122
– chemical structure of 7

Lewis acids 84
LLM-105 164
LLM-105 based PBX 164
Low molecular weight plasticizers 171
Low-vulnerability ammunition (LOVA) 105
LUMO (lowest unoccupied molecular orbital) 200
LX-07 163
– detonation properties of 163
LX-10 163
– detonation properties of 163
LX-17 163

Macroscale simulations 206–7
Manser, G.E. 10
MAPO 3
Mass spectrometry (MS) 43
Mathematical theory of homogenization (MTH) 207
Melt-cast explosives based on ETPE 141–42
Melt-cast formulations 141–42
Mesoscale simulations 205–6
Methyl-3-hydroxymethyl oxetane (MHMO) 87
3-methyl-3'-(tosyloxymethyl) oxetane (MTMO) 24
Methylene bis(cyclohexyl isocyanate) (MCHI) 31

Model potential/variable retention of diatomic differential overlap (MODPOT/VRDDO) 201
Molecular dynamics (MD) simulations 201–5
Molybdenum/vanadium oxide catalysts (MOVO) 58
MRD-CI (multiple reference double excitation–configuration interaction) 201

Nanothermite 144
Naval Surface Warfare Center (NSWC) 8
NIMMO-THF-NIMMO ETPE 130
Nitramidine 2
Nitramine oxidizers 50
Nitramine/ETPE-TNT melt-cast formulations 142
Nitramine-based CMDB propellants 102
Nitramines
– and mixtures with GAP 50
– thermal decomposition of 50
Nitrate ester oligomers, as energetic plasticizers 184–85
Nitrate ester plasticizers 180–84
– chemical structures of 180
– energetic parameters for 180
– energy parameters of 180
– general characteristics 180–83
– performance of 183–84
Nitrated cyclodextrin polymers (CDN) 89–90
Nitrated hydroxy terminated polybutadiene (NHTPB) 89
Nitrate-ester polymers
– combustion of 99–110
– – composite modified double-base (CMDB) propellants 100–5
– – NC based double-base propellants 99–100
– – PGN based composite propellants 108–110
– – Poly(NIMMO)-based composite propellants 105–8
3-nitratomethyl-3-methyl oxetane (NIMMO) 10, 87
– copolymers of 132
– polymerization of 88
– preparation of 87
Nitratoethylnitramine (NENA) 188–89
– properties of 188
– structure of 188
2-nitratoethyl oxirane 86
– structure of 86
Nitrocellulose (NC) 1–2, 13, 81
– chemical structure of 1

– double-base (DB) propellants, combustion of 99–100
– preparation of 81–2, 98
– thermal decomposition behavior of 90–3
Nitroglycerin (NG) 8, 81, 180
Nitro-groups based plasticizers 186–89
Nitropolymers 81–114
– combustion of nitrate ester polymers and propellants 99–110
– – composite modified double-base (CMDB) 100–5
– – NC based double-base DB propellants 99–100
– – PGN based composite propellants 108–10
– – poly(NIMMO)-based composite propellants 105–8
– explosive formulations 110–14
– preparation of 81–90
– – nitrated cyclodextrin polymers 89–90
– – nitrated NHTB 89
– – nitrocellulose 81–82
– – poly(glycidyl nitrate) 82–86
– – polynitratomethyl-methyl oxetane 87–88
– thermal decomposition behavior of 90–98
– – nitrocellulose 90–93
– – poly(glycidyl nitrate) 97–98
– – Poly(NIMMO) 93–97
Nobel, Alfred 2

Oligomers, of nitropolymers 185
One dimensional time to explosion (ODTX) 164
One-shot method, of polyurethane network formation 27
Optical microscopy (OM) 52
Organotin catalysts 30
Oxetanes, azido polymers from 22–24

Patrick, Joseph C. 2
PAX-2A 188
PAX-3 188
PBX 9205 6
PBX 9502 163
– detonation properties of 163
PBX 9503 163
– detonation properties of 163
PBX formulations 111
PBX formulations with fluoropolymers 162–64
PBX, interfacial debonding of 206
PBX-9404 110
– performance of

– – comparison with Comp B and Octol 110
– sensitivity parameters of
– – comparison with Comp B 110
PBXN-110, performance and shock sensitivity parameters of 111
PBXW 7 insensitive booster composition 164
PBXW 7 Type I
– detonation properties of 163
PBXW-115 65
Pelouze, Theophile Jules 1
Peroxide and amine cured vulcanizates
– comparison of 157
Peroxide-cure reactions 157
PETKAA (pentaerythritol tetrakis (azidoacetate)) 172
– chemical structures of 188
PGN/AN/energetic plasticizer propellants 108
Phase transfer catalyst (PTC) 20
Picatinny Arsenal (PAX series) 188
Plastic bonded explosives (PBX) 202
Plasticizer
– migration 185
– non-energetic 171
– oligomeric 174
Poly CDN 89–90
Poly(2-fluoro-2,2-dinitroethyl) polynitroorthocarbonate prepolymers
– general structure of 161
Poly(2-methyl-5-vinyl tetrazole) (PMVT) 67
Poly(2-nitratoethyl oxirane) 86
Poly(3,3-bis(azidomethyl) oxetane) (Poly (BAMO)) 197
– and copolymers with THF 198
Poly(AMMO) 24
Poly(BAMO) 11, 23, 172
Poly(butylene adipate) (PBA) 7
Poly(glycidyl nitrate) (PGN)
– based propellant formulations 108
– preparation of 82–86
– thermal decomposition behavior of 97–98
PGN Plasticizers
– advantages 185
– migration rates 185
PGN/AN/nitramine formulations 109
PGN/HMX/FOX-7 formulation 114
– comparison with Comp B 113
Poly(NIMMO) 10, 172, 185, 202
Poly(NIMMO)/ADN propellants 106, 107
Poly(NIMMO)-based composite propellants 105–8

Poly(tetrafluoroethylene) (PTFE) 147, 148–52
– energetic composites of 149–52
– phase transitions of 148–49
Poly(vinyl tetrazole) (PVT) 66
Poly(vinylidene difluoride) (PVDF) 202
Polyacrylonitrile (PAN) 66
Polybutadiene–acrylic acid copolymer (PBAA) 3, 13
– chemical structure of 3
Polybutadiene–acrylic acid–acrylonitrile copolymer 3
– chemical structure of 3
Polybutadiene–acrylic acid–acrylonitrile copolymer (PBAN) 13
Polybutadienes (PBAA, PBAN and CTPB) 3–4
Polychlorotrifluoroethylene (PCTFE) 202
Polydimethylsiloxane (PDMS) 201
Polyepichlorohydrin (PECH) 9, 19
Polyethylene glycol (PEG) 67
Polyglycidyl nitrate (PGN) 8–9
– chemical structure of 8
Polyisocyanates 25–36
– preparation of 26–28
Polymer blend 68
Polymer bonded explosives (PBX) 6
Polymer nanocomposites (PNC) 142–44
Polymer salts 68
Polymers, in binder applications 13
Polynitratomethyl-methyl oxetane (Poly (NIMMO))
– Poly(NIMMO)-based composite propellants 105–8
– preparation of 87–88
– thermal decomposition behavior of 93–97
Polynitrofluoroformal based energetic polymer 162
Polynitrofluoroformals
– randomly distributed copolymers of 162
Polyol initiators 84
Polyphosphazenes 11
– chemical structure of 11
Polysulfide elastomer 13
Polysulfide polymer (LP-3) 3
Polysulfides 2–3
– chemical structure of 2
Polyurethane elastomer 27
– schematic representation of 28
– structure of 28–9
Polyurethane network
– preparation of 26
Polyurethanes 4–5, 13
– chemical structure of 4

Prepolymer method
– curing reaction using 26
– of polyurethane network formation 26
Primary hydroxyl groups 21
Propellant formulation 3

Quantum mechanical methods 197–201, *200*
Quasi living cationic polymerization 24
Quasi-continuum approaches 197

R-45M 5
Rarefactions 154
Ratchet growth 205
RDX 101
ReaxFF 203
Representative volume element (RVE) 197
RF-67-43 explosive 111
– comparison with PBXN-110 *111*
Rocket equation 61
Rubber like polymer binders 6

Schonbein, Christian Friedrich 1
Screw extrusion technology 6
Shock wave energy 65
SMATCH (simultaneous mass and temperature change)–FTIR 41
Sodium azide (NaN$_3$) 19
N-stannylurethane 30
Styrene–ethylene/butylene–styrene (SEBS) 7

TATB (triaminotrinitrobenzene) 202
TDI (toluene 2,4-diisocyanate) 27
Tetrafluoroethylene, copolymers of 152–59
Tetrahydrofuran (THF) 197
Tetrazole polymer 67
– and their salts 66–69
TEX (4,10-dinitro-2,6,8,12-tetraoxa-4,10-diazaisowurtzitane) 51
Thermogravimetric analysis (TGA) 178, 188
Thermogravimetry (TG) 43
Thermoplastic elastomers (TPE) 6–8, 23, 121–22
Thiokol 3
T-jump–FTIR 41
TMNTA 172
– chemical structures of *188*
TNAZ (1,3,3-trinitroazetidine) 184
TNT (trinitrotoluene) 65
Triaminoguanidine nitrate (TAGN) 58, 109
Triaminotrinitrobenzene (TATB) 6
Triethylene glycol dinitrate (TEGDN) 180
Trimethylol ethane trinitrate (TMETN) 180
Triol 5
Triphenyl bismuth (TPB) 30

tris(azido acetoxy methyl) propane (TAAMP) 179
Tsiolkovsky, Konstantin E. 61

Vacuum thermal stability (VTS) 60
VDF-based fluoropolymers 157
Velocity of detonation (VOD) 63
Vibrational frequencies, of polymers 200
Vieille, Paul 2
Vinyl tetrazole polymers 67

Viton 147
– chemical structure of *156*
Viton A 156–59, 163
– thermal decomposition of 158–59
Viton A PBX 164
Viton curing, with a diamine curing 157

Xyloidine 2

Zirconium based formulations 103